GONE TOMORROW

GONE TOMORROW

The Hidden Life of Garbage

HEATHER ROGERS

THE NEW PRESS

NEW YORK
LONDON

Published in the United States by The New Press, New York, 2005
Distributed by Perseus Distribution

ISBN 978-1-59558-120-4 (pb.)
ISBN 978-1-59558-572-1 (e-book)

LIBRARY OF CONGRESS CATALOGING-IN-PUBLICATION DATA

Rogers, Heather, 1970–
Gone tomorrow : the hidden life of garbage / Heather Rogers.
p.cm.
Includes bibliographical references and index.
ISBN 978-1-56584-879-5 (hc.)
1. Refuse and refuse disposal--United States--History. I Title.
HD4483.R64 2005
363.72'85'0973--dc22

2005041562

The New Press publishes books that promote and enrich public discussion and understanding of the issues vital to our democracy and to a more equitable world. These books are made possible by the enthusiasm of our readers; the support of a committed group of donors, large and small; the collaboration of our many partners in the independent media and the not-for-profit sector; booksellers, who often hand-sell New Press books; librarians; and above all by our authors.

www.thenewpress.com

Book design and composition by Hiatt & Dragon, San Francisco

This book is printed on recycled paper.

Printed in the United States of America

10 9 8 7 6

This book is dedicated to those
who live and work with garbage

Contents

Acknowledgments

I began investigating the subject of garbage by making a documentary film in 2002, also titled *Gone Tomorrow: The Hidden Life of Garbage*. Once the film was complete, I realized that there was much more to say on the subject: thus this book. The people who contributed to the book overlap with those who helped on the film.

The insights of the interviewees in the documentary have been greatly influential: Tim Krupnick from the Batcave, Dave Williamson of Berkeley's Ecology Center, and Charles Homes all helped me more fully understand the issues and contradictions of recycling; Mary Lou VanDeventer offered an astute take on garbage and made key reading recommendations that set me on the right course; John Marshall continued to talk with me about labor and garbage well after the film was complete, and raised key points during the writing of the book; and, with his keen analysis, Richard Walker helped me grasp the complex meanings of garbage in a market economy.

In New York, thanks to the Brecht Forum, and to Cathryn Swan and Christina Salvi of Recycle This! for screening the movie and providing a context to discuss my ideas with a diverse range of people. Thanks as well to Robin Nagle for sharing some of her contacts at the New York Department of Sanitation and the Newark, New Jersey, incinerator.

I am indebted to David Harvey, who let me sit in on many of his classes while I was writing this book. His superb discussion of political economy deeply informs my analysis. Thanks as well to Christian Parenti for his intellectual comradeship, help on the film, introduction to The New Press, and early encouragement to write this book. Christian's observations and thoughts were crucial to this undertaking; it was with him that I most thoroughly discussed and worked through my ideas about garbage and how to write about it.

I am grateful to Colin Robinson for his overall support, engagement with the project, and absolutely crucial feedback on the initial manuscript. Liza Featherstone's enthusiasm for the movie and then the book has been invaluable. Once the writing was under way, she offered an abundance of useful comments, initially as a fellow author, and later as my editor—the book benefited greatly from her input. David Morris generously read an early version of the manuscript and made suggestions that proved extremely helpful. Praise to Steven Hiatt and Elinor Pravda for their careful, top-notch work on the production of the book. And much appreciation to Elizabeth Seidlin-Bernstein and The New Press staff and interns for all their behind-the-scenes work.

Thanks to librarians, who keep information free and accessible, and thanks to all the friends who sent articles and recommended books to read. For their help in providing photographs for the book,

I am grateful to Michael Lorenzini at the New York City Municipal Archive; Marguerite Lavin at the Museum of the City of New York; and James Martin, a retired public works engineer from Fresno, California. Thanks also to the artist Peter Garfield for his excellent portrait of garbage. Todd Chandler's stellar computer skills made organizing the photographs for publication a breeze. Thanks as well to Sam Cullman for additional technical assistance.

Much gratitude, for a variety of reasons, to Rachael Rakes, Scott Fleming, Penny Lewis, Emily De Voti, Tedd Hamm, Christopher D. Cook, Theresa Kimm, Thomas Green, Braden King, Susan Parenti, Josh W. Mason, Doug Henwood, Rick Prelinger, Neil Smith, Nancy Newhauser, and the Blue Mountain Center, where I worked out my initial ideas for the book. And warm thanks to my cousin Charlie Rogers, and my sister Holli Rogers, both of whom read parts of the book in a much rougher form and offered unswerving encouragement along the way.

Partial view of Fresh Kills landfill. (Courtesy of the New York City Department of Sanitation, Bureau of Waste Disposal)

Introduction

The Conquest of Garbage

> A society in which consumption has to be artificially stimulated in order to keep production going is a society founded on trash and waste, and such a society is a house built upon sand.
>
> DOROTHY L. SAYERS
> "Why Work?" 1942

From outer space several human-made objects are visible on earth: the Great Wall of China, the pyramids, and, on the southwestern tip of New York City, another monument to civilization, Fresh Kills Landfill. Briefly a depository for the gory debris of 9/11, this colossal waste heap looks rather like a misplaced Western butte. Its fifty-three years' worth of refuse are mostly covered by graded dirt and grasses, and not far off one can see what looks like a functioning estuary. On a bad day the methane stench of consumption past wafts up from the guts of the hill, and when storms hit, toxic leachate flows into the surrounding surface and groundwater. If garbage were a nation, this would be its capital. It's an astounding place, but apart from its size, not so unusual.[1]

In 2003 Americans threw out almost 500 billion pounds of paper, glass, plastic, wood, food, metal, clothing, dead electronics and other refuse.[2] Every day a phantasmagoric rush of spent, used and broken riches flows through our homes, offices and cars, and from there is burned, dumped at sea, or more often buried under a civilized veil of dirt and grass seed. The United States is the world's number one producer of garbage: we consume 30 percent of the planet's resources and produce 30 percent of all its wastes. But we are home to just 4 percent of the global population. Recent figures show that every American discards over 1,600 pounds of rubbish a year—more than 4.5 pounds per person per day.[3] And over the past generation our mountains of waste have doubled.[4]

Eat a take-out meal, buy a pair of shoes, read a newspaper, and you're soon faced with a bewildering amount of trash. And forget trying to fix a broken toaster, malfunctioning cell phone or frozen VCR—nowadays it's less expensive to toss the old one and purchase a brand-new replacement. Many people feel guilty about their waste and helpless over how to avoid it. This angst intensifies if our discards aren't promptly hauled away. Consider this tortured passage from a Chicago journalist writing during a 2003 garbage strike:

> I want to improve the environment. I do. But looking at an extra week's worth of Lucky Charms With New Larger Marshmallows boxes and Conair packaging and empty water bottles and ripped-up Instyle magazines and dried-out nail polish in Bus Stop Crimson and Gap bags. . . . I wonder if maybe I'm not doing my part.
>
> The mounting trash is a constant reminder of how much we spend. How much we consume. How much we waste.
>
> Won't someone please come take it away?[5]

Disruptions in the channeling of trash out of our immediate lives are infrequent; most often the rubbish gets collected on time. But even when the system works well, a nagging feeling can linger: Where does it all go? The opening lines of the popular 1989 film *Sex, Lies, and Videotape* bring these repressed anxieties to the surface. Andie MacDowell's character confesses to her therapist: "Garbage. All I've been thinking about all week is garbage. . . . I mean, we've got so much of it. You know, I mean, we have to run out of places to put this stuff, eventually." Many people today feel at least a little uneasy about the profusion of garbage that our society produces. But while its fate most often lies hidden, trash remains a distressing element of daily life, linked to much larger questions that never really go away.

Garbage is the text in which abundance is overwritten by decay and filth: natural substances rot next to art images on discarded plastic packaging; objects of superb design—the spent lightbulb or battery—lie among sanitary napkins and rancid meat scraps. Rubbish is also a border separating the clean and useful from the unclean and dangerous. And trash is the visible interface between everyday life and the deep, often abstract horrors of ecological crisis. Through waste we can read the logic of industrial society's relationship to nature and human labor. Here it is, all at once, all mixed together: work, nature, land, production, consumption, the past and the future. And in garbage we find material proof that there is no plan for stewarding the earth, that resources are not being conserved, that waste and destruction are the necessary analogues of consumer society.

This book focuses on household waste, or what is often referred to as "municipal solid waste," which includes rejectamenta from kitchens, bathrooms, hotels, schools, local shops and offices, and small construction sites. Even though over seventy tons of industrial debris from mining, agriculture, manufacturing, and petrochemical production are created for every ton of household discards, it is the slough of daily life that affects average people most directly because it is the waste we make.

Trash Ecology

The direct environmental impacts of garbage are sobering. Increased amounts of trash mean more collection trucks on the road spewing diesel exhaust into the atmosphere. Incinerators release toxics into the air and spawn ash that can contaminate soil and water. Landfills metastasize like cancer across the countryside, leaching their hazardous brew into nearby groundwater, unleashing untold environmental problems for future generations.

Garbage graveyards have met stricter environmental controls for only the past two decades, with a national set of standards being implemented just over ten years ago. This means that there are scores of dumps across the country still struggling to meet newer, tougher regulations. Meanwhile, they pollute local aquifers, soil and air.

In addition to leaching liquid wastes, landfills also erupt "landfill gas," the emissions of decomposing waste. This wretched vapor consists mostly of highly flammable methane, which is a major contributor to global climate change. According to the Environmental Protection Agency (EPA), "Methane is of particular concern because it is 21 times more effective at trapping heat in the atmosphere than carbon dioxide." Landfill gas also includes the air-borne wastes from things like adhesives, household cleaners, plastics and paints, including carbon dioxide, hazardous air pollutants (HAP) and volatile organic compounds (VOC). One EPA assessment reads: "Emissions of VOC contribute to ground-level ozone formation (smog). Ozone is capable of reducing or damaging vegetation growth as well as causing respiratory problems in humans . . . exposure to HAP can cause a variety of health problems such as cancerous illnesses, respiratory irritation, and central nervous system damage."[6]

Incinerators are just as onerous. As of the year 2000, according to the United Nations Environment Programme, municipal waste incinerators were responsible for creating 69 percent of worldwide dioxin emissions. Dioxins are a group of chemicals identified as among the most toxic in existence.[7] Even if a facility is outfitted with all the latest filtration gear, dioxin cannot be destroyed or neutralized because it is generated through the very process of incineration. When everyday substances like paper and plastic are burned together they form dioxin, which either takes to the air or lingers in the remaining ash. Incinerators also spew acidic gases (which cause acid rain), particulate matter, carbon monoxide and mercury. Leftover ash can also contain heavy metals like lead, mercury, cadmium and other toxic substances that can leach once buried in landfills.

Packaging comprises the largest, and most rapidly growing, category of discards. More than 30 percent of municipal waste is packaging, and 40 percent of that is plastic. Polymers are now ubiquitous in the country's incinerators and landfills as they overflow with televisions, computers, cell phones, medical waste, soda bottles and cellophane wrappers.[8] But because they are resilient and toxic, synthetics cannot be safely returned to the environment. On average, plastics are predicted to stay intact for 200, 400, maybe 1,000 years, and these are only guesses.[9] For their life span, plastics kept above ground will abrade and "off-gas" malignant releases into the air.[10] When buried, plastic resins can leach hazardous materials into the water and soil, and during production polymers are notoriously dangerous, poisoning workers and the environment. Regardless of these problems, the U.S. plastics industry has boomed over the last fifty years, growing at twice the annual rate of all other manufacturing combined.[11] Perhaps that's why the middle of the Pacific Ocean is now six times more abundant with plastic waste than with zooplankton.[12]

Recycling was presented as a solution to the garbage crisis, but it can't keep pace with the staggering output of throwaways. About 80 percent of U.S. products are used once and then discarded.[13] Although there are more than 9,000 curbside recycling programs in the country, many towns do not collect the stuff. And even if the dutiful separate their metal from glass, much of it still ends up at the landfill or incinerator, having found no buyer on the other end. If substances sent to recovery centers can't compete with lower-priced "virgin" materials, they get dumped. And, further limiting the expansion of recycling, U.S. producers are not required to use reprocessed materials even though most manufacturers now stamp their containers with the eco-friendly recycling symbol.

Perhaps surprisingly, 50 percent of all paper ends up as garbage (in fact, paper accounts for fully half of the discards in U.S. landfills).[14] Today, only 5 percent of all plastic is recycled, while almost two-thirds of all glass containers and half of aluminum beverage cans get trashed.[15] Tossed as soon as it is empty, sometimes within minutes of purchase, packaging is garbage waiting to happen.

There are other catastrophic environmental consequences of a mass consumption society. While they may not appear explicitly connected to garbage, these effects are unmistakably bound with a system that produces so much trash. Just as manufacturing creates the bulk of all refuse, the mass production that creates the torrent of cans, bottles, and electronics we throw away wreaks havoc on nature due to the insatiable need for raw materials. Extraction of timber, natural gas, oil and coal, along with massive water and power usage, all speed the destruction of the earth's natural systems.

Forecast: Extreme Conditions

The rate of climate change is perhaps the broadest barometer of environmental health and is closely linked to trash; the more that gets thrown out, the more pollution-causing processes are relied on to make replacement goods. According to the latest reports, global warming is increasing at a far faster rate than previously predicted. In contravention of the Kyoto Protocol (which the U.S. refuses to sign), America's carbon emissions from burning fossil fuels—a major contributor to heat-trapping greenhouse gases—increased almost 20 percent between 1990 and 2000.[16]

Most likely the result of these emissions, extreme weather has already caused a series of apocalyptic scenarios. A 2003 heat wave in France killed 15,000. A quartet of hurricanes in 2004 slammed into the Caribbean and Florida, slaying hundreds, displacing thousands and causing an estimated $25 billion in damages in Florida alone. On the other side of the planet, Bangladesh suffered the most violent flooding in a half-century during the summer of 2004. Almost 800 people drowned, 30 million were left homeless and countless crops were ruined. The storms left behind more than $6 billion in damages to roads, agriculture and industry.[17]

Not only are the knock-on environmental effects exported from developed countries to the global South, so are mass production, consumption and wasting patterns. As factories move overseas seeking ever-cheaper armies of labor, they disseminate manufacturing practices that churn out epic amounts of trash and pollution, beyond the reach of the global North's tighter environmental controls.

Consumption has been transformed in places like India, where ceramic throwaway teacups are being replaced by disposable plastic cups and bottles. Instead of getting harmlessly ground into the roadside, polluting plastic must be buried or burned. Plastic shopping bags are unleashing their wrath across developing countries as well. In China they're known as the "white pollution," while South Africans call trashed polymer bags the "national flower." India is flocked with the soft, thin material; it hangs from trees, litters streets, ruins rivers and chokes sacred cows that consume it while grazing. And in Bangladesh, plastic bags have clogged and destroyed drainage systems, causing such major flooding that the government outlawed the manufacture of disposable synthetic totes in 2002.[18]

Adding still more to these new and growing wellsprings of trash, much of America's discards get shipped overseas for recycling and disposal. These encompass used plastic bottles, scrap metal, spent chemicals, and a virulent new category of refuse that includes trashed televisions, computers, cell phones and other electronics, known as "e-waste." And foreign processing firms are buying rubbish from the United States—the world's largest consumer—at a rapidly expanding rate. Exporting wastes from the United States, an industry worth just under $200 million in 1997, grew to over $1 billion in revenues in 2002.[19] Container ships from Asia that bring brand-new sneakers, Teflon kitchenware and CD players to the United States now return home packed with American rubbish.

How did we get into this mess? Consumption lies at the heart of American life and economic health, and intrinsic to consumption is garbage. Such high levels of waste are the product not of any natural law or strange primordial impulse but of history, of social forces.

The world of trash did not always exist as it does today. In the nineteenth century refuse was sorted, municipal waste was composted, and all kinds of materials that left the home as discards were extensively reused. But with industrialization and two massive world wars the production system was radically transformed, and so too was garbage. Increasingly, what gets thrown away is shaped by monopolistic corporate power: at one end manufacturers, marketers and ad men, at the other giant corporations like Waste Management Inc. With annual consumer spending in the United States now accounting for about two-thirds of the nation's $11 trillion economy while outlays on discard handling and disposal approaches $50 billion, it's no wonder there's so much trash—garbage is good for business.[20]

Household garbage being processed at a landfill. (Lawrence Racioppo, 2000)

1

The "Waste Stream"

Most Americans set their full garbage cans out on trash night and retrieve them empty the next morning. Aside from fleeting encounters—such as a glimpse of a collection truck trundling down a neighborhood street—many people have only a vague sense of where their discards go. They may think that trash is benignly and permanently disposed of in "proper" places. However, the truth is that these sites are filthy, mysterious and ultimately only short-term palliatives. The veiled and dirty story of the discard treatment process tells us more than just where our effluent goes; it also reveals the vast resources channeled into getting rid of our dead commodities.

While for some householders the bulging trashcan placed on the curb might seem like the end of the line, it actually marks the beginning of what many municipal sanitation departments now euphemistically call the "waste stream." Seemingly imperceptibly, this flow of refuse gets processed through a lengthy and complex system that is grisly, oddly fascinating, and integral to the functioning of our daily lives and the metabolism of the market.

Transferred

In the dark chill of early morning, heavy steel garbage trucks chug and creep along neighborhood collection routes. A worker empties the contents of each household's waste bin into the truck's rear compaction unit. Hydraulic compressors immediately scoop in and crush the dross, cramming it into the enclosed hull. Since collection is the most expensive part of the refuse treatment process, condensing the stuff makes for more efficient use of valuable hauling space, but also means that potentially reusable objects are often immediately destroyed and rendered unsalvageable. When their rigs are full, the collectors return to a garbage depot called a transfer station, where they unload. Once empty, the trucks head out for the next round and continue on this cycle until the day's work is done.

The transfer station has been used in basically the same way since the earliest days of organized waste collection in the nineteenth century. There was always a limit on the amount a carter could haul in one load, so strategically located lots for temporary dumping were essential. As in the past, today's transfer stations tend to be huge warehouses, sheds and yards where tons of garbage is stored, sorted, consolidated and reshipped either by eighteen-wheelers, railcars or barges. Typically located in poor and working-class neighborhoods near highways, waterways and rail lines, these transfer stations stink and contaminate the soil. They are a nexus of nonstop diesel exhaust and house giant populations of rats and other pests.

San Francisco's main waste-processing hub is pretty typical for a city its size. This transfer station, owned by Norcal, a state-wide private rubbish handler, is located in the southern part of the city, nestled between a major freeway, several public housing projects and a working-class immigrant neighborhood. Most of the city's household wastes and a portion of its commercial discards are processed at this huge forty-four-acre facility.[1]

Sprawling across the site are several low industrial buildings connected by dirt access roads. Compactor garbage trucks stream in and out of a cavernous pitched-roof warehouse, referred to by Norcal employees as "The Pit." Here plastic garbage bags are disgorged from the hydraulic collection trucks by the tens of thousands, their snagged and ruptured bladders spilling tons of waste down into a football-field-sized concrete canyon full of rotting refuse. Swarms of screeching seagulls hover overhead.

Looking down into this hellish scene, one sees a bulldozer, half-submerged in the goo, splattered and caked in muck, its windshield covered in a thick translucent film that obscures the lone driver. Back and forth the dozer grinds and lurches through the garbage, compacting, compressing, and moving oversized heaps away from the zone where the trucks unload. From inside The Pit's high metal roof falls a steady mist of water to keep the dust and paper down. The stench is unbearable: fetid, greasy, cloying, it penetrates one's clothes and skin, lingering persistently in the nose and mouth. Garbage workers say they can never fully wash the smell away. Toward the end of the day, the dozer's eight-foot steel spade shovels this stewing, compacted mess into cargo containers that are hitched to trucks and hauled about fifty miles east to the company's Altamont landfill. (Other transfer stations might send their refuse to an incinerator.)

On the west side of the compound in another titanic corrugated steel structure is the Materials Recovery Facility, or MRF.[2] This is where one finds a more inspiring side of garbage: recycling. Here glass, plastic and metal get dumped from collection trucks onto a wide conveyor belt, which moves the materials through a series of mechanical sorting operations. At one section, blasts of pressurized air blow light plastic off the belt, leaving the heavier metal and glass behind. The plastic is then sucked through a huge overhead tube into a separate part of the warehouse—a cavernous, otherwise empty room heaped twenty feet high with spent milk jugs, water bottles and yogurt containers.

At another point, recyclables flow onto a special perforated conveyor belt that shakes and jerks to sift out broken glass, which falls through the holes into a receptacle below. (The broken glass is sold separately for less money than whole bottles.) Then the materials pass a series of powerful magnets that attract ferrous metals, leaving mostly the valuable and easy-to-sell aluminum behind. (Aluminum is the only item in our current waste stream that is profitably and regularly recycled.) Workers oversee the entire process, checking for contamination and sorting what the machines cannot, like differentiating green from clear or amber glass. At the end of this complex conveyor system, each branch of the belt drops the sorted material into the appropriate receptacle.

Midway between Norcal's Pit and its MRF in yet another concrete slab metal warehouse is the paper sorting and baling station. Here workers direct the bulging collection trucks as they off-load newspapers, junk mail, office paper and magazines bundled in brown grocery bags. These longer, partially open, split-bin trucks do not compact their cargo like garbage haulers and so slough the payloads from their interiors in loose and slippery heaps; the air hangs thick with a strange paper dust. From there a tractor gathers the material and dumps it into a baler, which loudly and lethargically excretes giant blocks of compressed printed matter. Workers driving fork-lifts then organize these easily transportable units into orderly rows. From here the bales of paper will be loaded onto another truck and, like the other salvaged materials, hauled either to a recycling center or more likely to a broker for resale.

The gentle slope of a grassy hill rises over the activity at Norcal's rambling facility. Its south-facing side is encrusted like a giant mosaic with items employees have plucked from oblivion. Matted stuffed animals frolic with sun-faded plastic gnomes. Toys, old Christmas decorations, and unusual items like a cowboy hat, an antique lamp and a sunken disco ball fill out the ever-expanding composition. It's a bizarre site in this otherwise highly circumscribed zone, where the most wretched mess is always well contained. This installation pays homage to all those commodities that disappear when they get thrown away. It resembles a spontaneous altar, the kind that might form on a street corner where someone was killed in a car accident. It is a form of folk art that's simply a human response to witnessing so much waste.

Out of Sight

Once garbage leaves the transfer station it goes either to an incinerator or, more likely, to what's called a "sanitary landfill." Encapsulating garbage in sealed underground plastic "cells," the newest landfills are expected to hold their densely compacted trash in perpetuity. Unlike their precursors, today's most advanced land disposal sites are meticulously engineered and closely regulated and have monitoring systems for wastewater and airborne emissions. And, like so much else in American culture, today's fills are supersized—new facilities are referred to as "mega-fills." The subterranean cells that comprise mega-fills range in size from ten to one hundred acres across and up to hundreds of feet deep. One Virginia whopper has disposal capacity equivalent to the length of one thousand football fields and the height of the Washington Monument.[3] Costing on average $500,000 per acre for research, development, and construction, these new garbage graveyards are awesome, eerie scenes.

There's a reason landfills are tucked away, on the edge of town, in otherwise un-traveled terrain, camouflaged by hydroseeded, neatly tiered slopes. If people saw what happened to their waste, lived with the stench, witnessed the scale of destruction, they might start asking difficult questions. Sitting atop Waste Management Inc.'s Geological Reclamation Operations and Waste Systems (GROWS) landfill, a large hill that rises three hundred feet composed entirely of garbage, the logic of so much consuming and wasting quickly unravels.[4] The fill's "working face," where the active dumping takes place, is a massive thirty-acre nightmare. A bizarre and filthy enterprise, the scene is populated by clusters of trailer trucks, yellow earthmovers, compacting machines, steamrollers and water trucks. They churn in slow motion through this surreal landscape, remaking the earth in the image of garbage. Throngs of seagulls hover, then plunge into the rotting piles, the ground underfoot is torn from metal treads, and potato chip wrappers and spare tires poke through the dark dirt as if floating to the surface. The smell is sickly and sour.

The aptly named GROWS landfill is part of WMI's 6,000-acre trash megafacility just outside Morrisville, Pennsylvania. As of 2002, GROWS was the single largest importer of New York City's garbage and one of Pennsylvania's biggest landfills. Although refuse from New York and other metropolitan centers gets hauled to Virginia, Ohio and other states, Pennsylvania is the country's leading rubbish importer.[5] The sprawling WMI compound also includes a waste incinerator, another newer landfill, a recycling center, a leaf composting lot, two soil mines and a reconstituted wildlife area that the company can't stop boasting about.

Perched on the banks of the Delaware River, this land has long served the interests of industry. Overlooking a sprawling, mostly decommissioned U.S. Steel plant, WMI, the world's largest trash conglomerate, now occupies the former grounds of the Warner Company. In the previous century, Warner surface-mined the area for gravel and sand, much of which was shipped to its cement factory in Philadelphia. The area has since been converted into a reverse mine of sorts; instead of extraction, scores of workers dump, pack and fill the earth with almost 40 million pounds of municipal wastes daily.

Back atop the GROWS landfill, twenty-ton dump trucks gather at the low end of the working face, where they discharge their fetid cargo. Several feet up a dirt bank, a string of larger trailers gets detached from semi trucks. In rapid succession each container is tipped almost vertical by a giant hydraulic lift and, within seconds, twenty-four tons of rotting garbage cascades down into the day's menacing valley of trash. A "landfill compactor," which looks like a bulldozer on steroids with mammoth metal spiked wheels, pitches back and forth, its fifty tons crushing the debris into the earth. Another smaller vehicle called a "track loader" maneuvers on tank treads, channeling the castoffs from kitchens and offices into the compactor's path. The place runs like a well-oiled machine, with only a handful of workers orchestrating the burial.

Move a few hundred yards from the landfill's working face and it's hard to smell the rot or see the debris. The place is kept tidy with the help of thirty-five-foot-tall fencing made of "litter netting" that surrounds the perimeter of the site's two landfills. As a backup measure, teams of "paper pickers" constantly patrol the area to snatch up any stray trash. Misters dot fence tops, roads and hillsides, spraying a fine, invisible chemical-water mixture into the air, which binds with odor molecules and pulls them to the ground.

In new state-of-the-art landfills, the cells that contain the trash are built on top of what is called a "liner." The liner creates a giant underground bladder that is intended to prevent contamination of groundwater by collecting leachate—liquid wastes and the rainwater that seeps through buried trash—and channeling it to nearby water treatment facilities. WMI's two Morrisville landfills leach on average 100,000 gallons daily. If this toxic liquid contaminated the site's groundwater it would be devastating.

Liners are constructed differently in varying terrains and climates, but in wetter regions generally look like this: several feet of earth are hard-packed; then a half-inch layer of bentonite clay padding

("Claymax") is laid down. Next comes 60-mm-gauge black plastic sheeting, made of high-density polyethylene (HDPE, a thicker version of the material used for milk and detergent jugs); an inch-thick plastic drainage mesh is then installed; on top of that goes another layer of bentonite padding and 60-mm HDPE; then a half-inch-thick synthetic felt fabric gets laid down to protect the layers underneath. Topping it all off is eighteen inches of gravel to facilitate drainage.

Once the cell is filled with trash, which might take years, it is closed off or "capped." The capping process entails covering the garbage with several feet of dirt, which gets graded, then packed by steamrollers. After that, layers of Claymax, synthetic mesh, and plastic sheeting are draped across the top of the cell and joined with the bottom liner to fully encapsulate all those worn out shoes, dirty diapers, old TVs and discarded wrappers.

Capped landfills are subject to ongoing monitoring in order to comply with various state and federal regulations, many of which came about due to the Clean Air Act of 1970 and the Clean Water Act of 1972. The operators of high-tech fills regularly check leachate and groundwater for toxicity while they extract "landfill gas"—methane, carbon dioxide, volatile organic compounds, hazardous air pollutants, and odorous compounds released by decomposing garbage—through a series of wells. The wells are drilled into the capped fill and use a vacuum system to suck the vapors from the rotting refuse. If handled properly, the gases are either burned ("destroyed") or turned into electricity after they've been collected.

That's how the landfill runs if all goes well. Not surprisingly, improved landfills have their flaws. Employees and regulators have discovered pollutants in Virginia's mega-fills, often as a result of "cocktailing," the illegal mixing of restricted hazardous and toxic wastes with regular municipal discards. Medical waste, including blood, chemotherapy waste, biohazard bags, human body parts and radioactive castoffs, and industrial refuse like asbestos and lead paint have been found.[6] Groundwater contamination by heavy metals was detected at two of Virginia's mega-fills, raising questions about the reliability of these facilities.[7]

The new way of rubbish disposal also impacts ownership in the industry. Because building high-end garbage graveyards is so expensive, fewer small businesses and municipalities can afford to go into the trash trade. This leaves the field wide open for well-capitalized firms like the behemoth WMI. And because the overall number of disposal companies is shrinking, fewer firms now have more leverage in setting prices and more power influencing public policy.

It's not that the interests of the trash conglomerates are all that different from those of smaller companies; increased garbage equals bigger profits for both. The difference between the two lies in scale. In Pennsylvania over the last decade the number of garbage disposal sites has shrunk from several hundred to about fifty, but disposal capacity has soared. That means far fewer companies now own bigger sites and—crucially—given the ability to bury or burn more trash, they do. Previously, there were logistical constraints on small companies; with less access to capital they could grow only so fast and they could consequently dispose of only so much garbage. Today, with large trash corporations, the limits are sky-high.

The giant rubbish firms also hold considerably greater sway than their quaint predecessors over political decision making. In a recent development on the policy front, the EPA under industry-friendly George W. Bush has proposed deregulating landfills nationwide, arguing that mandated environmental protections have become obstacles to innovation.[8]

Barring such an erosion of protections, today's requirements, ranging from liner construction to post-capping oversight, mean that disposal areas like WMI's GROWS are potentially less dangerous than the landfills of previous generations. But the fact remains that these systems are short-term solutions to the garbage problem. While they may not seem toxic now, all those underground cells packed with plastics, solvents, paints, batteries and other hazardous materials are time bombs that will someday have to be treated since the liners won't last forever: most liners are guaranteed for only fifty years or less. That time frame just happens to coincide with the post-closure liability landfill operators are subject to; thirty years after a site is shuttered, its owner is no longer responsible for contamination, the public is.[9]

Cleansing by Fire

Incineration is still the most contested and costly disposal method of all, receiving only about 15 percent of household waste in the United States today. Unpopular since its appearance in the nineteenth century, the incinerator inspires public opposition because it spews stinky, dangerous smoke, gases and ash into the air. Incinerators have also always been one of the most expensive treatment methods around; these capital-intensive plants require costly initial outlays, constant maintenance and larger numbers of skilled workers than do land dumps. And trash crematories are straight out of hell.

Camouflaged by a neutral big-box exterior, American Ref-Fuel's Newark, New Jersey, incinerator houses a macabre scene of nonstop destruction.[10] In the bowels of the plant is a vast ten-story coliseum that's more than two hundred feet long and fifty feet wide, filled with garbage. On this scale the individual items in each household trash bag become insignificant; so much refuse amassed in one place stuns the senses. This monolithic room is where all the rubbish is collected before it gets burned. After being unloaded from packer, trailer and dump trucks in an adjacent warehouse, refuse is plowed through any of sixteen passageways into this giant hull. The light is amber and dim, the air hazy. There are no visible flames, only massive concrete chutes through which the trash travels to its fiery fate. Hanging from the ceiling is a gargantuan steel grapple; its long talons can seize ten tons in one go. The grapple swoops down and sinks its spikes into the sea of trash below, then cranes up, dust and paper swirling in a trail behind it. The leviathan then pulls toward one of three concrete funnels and slowly, swaying, releases its prey.

Guiding the grapple is a lone operator in a small control room located just below the ceiling, overlooking the pit. The worker sits in pitch-black darkness, facing a large glass wall. He looks as if he's piloting a spaceship in an apocalyptic science fiction film. His battered but futuristic chair takes up much of the cramped chamber. Its tall black back is flanked by wide metal armrests dotted with control switches. Both hands clench joysticks, his movements telling the grapple where to go. He sits intently, his body rigid as he opens the pincers wide and reels them down to clutch a snarl of bursting plastic garbage bags. Carefully but quickly the operator lures up then releases the tons of broken appliances, torn clothing and rotting food into one of the fuel channels that feed the fires. Just above the operator's right shoulder, two closed-circuit televisions beam faded, slightly distorted black-and-white images from cameras positioned in the mouths of the chutes. The grapple controller uses these to monitor the flow. On the small screens one can see the backdraft spitting up paper scraps and lightweight debris from the belly of the burner as the mounds of discards slowly, unceremoniously sink into the flames.

Located about thirty miles west of New York City, this Essex County facility burns 2,800 tons of rubbish daily. Its "fuel" comes from the surrounding towns, including New York City, and sometimes includes such industrial substances as cases of excess pharmaceuticals, latex paint and expired Polaroid film. The incinerator operates seven days a week, twenty-four hours a day, releasing gas and smoke from its stack, which is almost three hundred feet tall. The ash the plant produces is buried at nearby landfills or sent farther afield to defunct Pennsylvania mines where it is used to stop up cavities left from decades of coal extraction. The heat from the furnaces is used to generate electricity.

Once the garbage is burned, the by-products are directed through an intricate treatment system. First, the heaviest ashes are collected in a cold-water trough. Then airborne gases, smoke, and the lighterweight "fly" ash that remain are channeled through a cooling system and into a series of high-tech filters called "scrubbers." The scrubbers are meant to take out pollutants like poisonous carbon monoxide and dioxin; sulfur dioxide and hydrochloric acid (both of which cause acid rain); and nitrogen oxide (which impairs the respiratory system, contributes to ground-level ozone, and also causes acid rain). After that, the flue gas and ash are directed through chambers that spray carbon molecules, lime slurry, and ammonia that attract and theoretically neutralize a range of virulent molecules. Finally, the airborne waste is sent through two giant metal plates charged with an electric current to attract fine particulate matter. The remaining material then leaves the stack and disperses into the air.

Armed with minute-by-minute computer data on a smattering of the plant's toxic releases and an annual third-party emissions audit, an American Ref-Fuel engineer asserts that the gas, smoke and ash that leave the building are safe. "Waste-to-energy is the most environmentally sound waste disposal method," he dutifully recites.

Set foot in any facility along the shores of the "waste stream" and one is confronted with a barrage of PR about how environmentally responsible today's methods are. "We're not trying to create pollution, we're trying to stop it," explains one earnest engineer at a midsized sanitary landfill in northeastern Pennsylvania.[11]

Two representatives at the WMI site that houses the GROWS fill are much more insistent about their love of nature. Traversing the facility's molested grounds in a mammoth white pickup, we pass a small field, which, they interrupt each other to explain, was planted with sunflowers to feed passing birds. These spokespeople are also eager to show off the site's Penn Warner Club, a camping, hunting and fishing area surreally located just yards from the operation's incinerator and twin landfills.

WMI's representatives boast that, on top of all this, they are engaged in educational activities in the surrounding community. Despite the fact that the WMI facility does not recycle e-waste—here those discards are burned or buried—the following day the firm is hosting an electronics recycling event in a nearby town. WMI also liaises with local Bucks County public schools to teach children to "put litter in its place." Last year the company produced a glossy calendar with area children featuring the theme slogan "Earth Day every day . . . just a clean-up away!" The high month of April—when Earth Day falls—is adorned by a drawing with prescriptions of where to tidy up each day of the week; Sundays are reserved for cleaning "my own yard." The messages scrawled in crayon communicate the now common refrain that garbage is each individual's responsibility.

It's not that any of these endeavors are bad. On the contrary, some of them bring real benefits, like the endangered turtle that's taken up residence in the Penn Warner Club's artificial lake. But precisely how helpful these "environmental" actions are in real terms remains unclear. The larger point is that firms like WMI exploit such activities to generate PR that camouflages the darker, indisputably more destructive side of their industry.

When asked why all these ecologically minded efforts are so important, the site's general manager responds from behind the wheel of his truck, the GROWS mound looming in the distance: "It shows that what we build, construct, design can be part of nature. Landfills aren't a bad thing. We have thousands of geese—see, there's our geese. I'm proud I can drive around and see a red-tailed hawk, we can have hunting and fishing. It must say we're doing a good job." If only the true environmental measure were so simple.

There is a palpable tension at waste treatment facilities, as though at any minute the visitor will uncover some illegal activity. But what's most striking at these places isn't what they might be hiding. It's what's in plain view. The lavish resources dedicated to destroying used commodities and making that obliteration acceptable, even "green," is what's so astounding. Each garbage collection system, transfer station, recycling center, landfill and incinerator is an expensive, complex enterprise that uses the latest methods developed and perfected at laboratories, universities and corporate campuses across the globe. An examination of this accumulation of technological innovation, scientific inquiry and geological and atmospheric study, coupled with unrelenting, disciplined public relations, community outreach and education, reveals much about this society's priorities.

The social, political and financial power that drives ever-more sophisticated attempts to annihilate discarded commodities is staggering. The more efficient, the more "environmentally responsible" the operation, the more the repressed question pushes to the surface: What if we didn't have so much trash to get rid of?

Even if all the PR-savvy flacks at waste treatment facilities are right that their processes are ecologically benign and will remain so, these places offer a view into the scope of what's possible. If it's feasible to create the kinds of facilities that handle our garbage today—to bond poisons with neutralizing materials at the molecular level as they're leaving an incinerator stack, to build a theoretically impermeable underground rubbish cell that can be monitored for hazardous gas and liquid releases—then surely it must be possible to restructure production to eliminate waste before it gets made.

The story of garbage encompasses two spheres: where it goes but also where it comes from. After all, refuse is widely perceived as simply a natural outcome of human existence; life inevitably begets rubbish. In some ways this is true—we excrete wastes from our bodies and garbage results from the most fundamental activities, like eating. But the way Americans waste today is not just a normal result of organic human development. It is the outcome of a long process rife with social, economic and political struggle.

Scow being unloaded into the ocean. (From *Collection and Disposal of Refuse* by Rudolph Hering and Samuel A. Greeley, 1921)

2

Rubbish Past

In 1841 in her *Treatise on Domestic Economy*, Catherine Beecher advised that "broken earthen and china can often be mended, by tying it up, and boiling it in milk. *Diamond cement*, when genuine, is very effectual for the same purpose. . . . A strong cement may be made, by heating together equal parts of white lead, glue, and the whites of eggs."[1] Another how-to book of that era suggested fixing broken glass with mixtures such as plaster of Paris and egg white, or just garlic juice, which, once applied, had to "remain undisturbed for a fortnight." And when the damage was beyond repair, household guide books recommended alternative uses for broken kitchenware: "If the dishes do not look well enough to come to the table, they will yet do to set away things in the store-closet, or for keeping jelly, marmalade, or preserves."[2]

This resourceful conservation of materials did not come from a sensitivity to the limits of nature's abundance, nor did it stem in most cases from a concern for the environment as it is conceived of today. Prudent consumption was directly linked to the availability and cost of manufactured goods; as long as commodities were expensive or difficult to obtain, they were tended and mended to last as long as possible. Thus, the primary discards in the eighteenth and early nineteenth centuries were organic discards—food scraps, manure and human wastes.

Since there was so much land in the earliest days of colonial farming, a grower could simply move to another plot when more fertile soil was needed, so organic castoffs often went unused. However, as garden markets expanded around major industrial cities, the farm's location became crucial, and wastes took on a whole new value. Throughout the first half of the nineteenth century, animal dung, human excrement, kitchen slop, street sweepings, and household wastes like ash were reused extensively by farmers to fertilize their fields. These discards were gathered from city streets, shops and houses to sell to growers for use as soil amendment, marking one of the earliest forms of waste collection.

Garbage changed dramatically after the Civil War as industrialization sped production and distribution, which drove down prices and concentrated ever-larger populations in urban centers. Manufactured items became dramatically cheaper and more accessible, leading to unprecedented levels and novel forms of waste. Also during these years, corrupt local officials failed to clean trash from city streets and left slums to steep in filth. Such appalling conditions, in turn, fueled epidemic disease and incendiary class conflict.

Typically, some plague, pestilence, or public upheaval would wreak havoc in communities, killing scores and setting off political and biological aftershocks. These forces of social breakdown, in turn, produced the everyday structures of social order. Catastrophe generated an organized and organizational response; urban planning, building codes, zoning, fire services, policing, courts, jails, public hospitals, schools, sewage treatment and basic sanitation all emerged in reaction to the multifaceted and interconnected crises of crime, disease and rebellion. Fearing for their lives and struggling to maintain social order, the nineteenth-century middle and upper classes organized to reform city government and improve health and cleanliness, helping lay the groundwork for modern municipal sanitation.

Before Garbage

Garbage as we know it is a relatively new invention predicated on the monumental technological and social changes wrought by industrialization. Until mass production became the norm in the United States, manufactured commodities were always expensive and not always available; most ready-mades were imported from Europe.[3] Such items were too dear to use once, then discard. Accompanying this economic fact was a rather different cultural logic than applies today. Even though colonial farms and cobblestone city alleys were cluttered with debris, the waste of preindustrial and early industrial societies was comparatively minimal and could for the most part be absorbed back into the earth. That's not to say there weren't serious health problems due to a lack of organized refuse management in cities, but the *contents* of the rubbish bin were relatively benign.[4] Traditionally, castoffs did not stand in opposition to nature so much as they were nature temporarily out of place.

During the seventeenth and eighteenth centuries most American settlers threw almost nothing away; they were so poor that manufactured goods were almost totally absent from their lives. According to colonial historian Alan Taylor, probate inventories of early farmers revealed that "common houses contained little furniture, usually only a bed, a table, a few benches, and a chest or two for clothing. The common people ate with their fingers, sharing a bowl and drinking from a common tankard, both passed around the table."[5] Under such spartan conditions, settlers were compelled to save fat for making candles and soap, and for cooking. Leftovers were boiled into soup or fed to chickens and hogs. Spoiled food wastes were tossed out windows to decompose in gardens. In colonial cities and towns, household practices were no different. Most people dumped discards like broken crockery and organic scraps in backyards and out windows and doors to the street, leaving the stinking clots of waste to rot untended or get eaten by roaming hogs, dogs and raccoons.[6]

"Night Soil" and Other Resources

In the times before garbage as we know it, one of the most voluminous of waste products was excrement, both animal and human. And it was in the treatment of dung and feces that the first systematic refuse collection developed. At the heart of the matter was the relationship between cities and the countryside; in a double-helix-like transformation, farming and urbanization dialectically shaped each other through waste handling. Simply stated, shit was the key to healthy soil just as it was the bane of cities.

The background story in cash-cropping colonial America was that soil practices were fundamentally off-kilter. From the earliest days of colonization, farmers ignored long-proven methods of maintaining soil fertility because in the New World land was seemingly inexhaustible, while labor—needed to feed, rebuild and conserve soil—was scarce. Instead of rotating crops and fertilizing with manure, settlers "mined" virgin soil for all it was worth and then moved on. "How scrupulously careful is the good husbandman of the produce of his farm . . . and yet how often careless of the food which can alone nourish and mature his plants!" wrote one exasperated, forward-thinking proto-agronomist.[7]

Beginning around 1800, after more than a century of this profligate land use, East Coast farmers, faced with exponentially declining yields, had to change their ways. Their soil was maxing out just as competition from farther west was ramping up and demand from urban markets like New York City and Philadelphia began to boom.[8] Since location was now key to distribution, farmers could no longer simply relocate once their current land was exhausted.

As a result, nineteenth-century cultivators began embracing Old World methods of applying wastes to the soil to boost production; after all, the more they could grow, the more they could sell. For example, the average farmer in Riverhead, New York, harvested sixty-four bushels of potatoes, but could harvest over a hundred if he used fertilizer.[9] Agriculturists initially focused on excrement as the best soil amendment. In 1833, *American Farmer* told its subscribers, "The importance of manure is such that we again allude to the subject, and would most particularly impress upon the minds of every cultivator, that manure is the grand moving power in the production of an abundant return.[10] So valuable had excrement become that one Brooklyn, New York, farmer stipulated in his will that his son should inherit "all manure on the farm at the time of my decease."[11] Soon a growing range of wastes was promoted as efficacious, as the editor of *Working Farmer* instructed: "save everything in the shape of refuse or offal, it is all good to make crops grow."[12]

If manure was the "grand moving power" on the farm, then hay was its urban corollary. Sold in the city by garden market farmers as animal food, hay was the chief fuel source for city horses, upon which the circulation of goods and people depended. By 1880, Manhattan had more than 150,000 horses trotting through its streets; the consumption of hay was therefore enormous, rivaled only by the unstoppable output of horse droppings.[13]

But with escalating demand from city dwellers for greater amounts of produce, farmers turned increasingly to external fertilizers to maintain the health of the soil. As the nineteenth century wore on, all variety of rejectamenta, including human feces, known among professionals as "night soil," became commodities to be bought and sold, much of it collected in urban centers and then hauled back to the farm. Either gathering the muck themselves or securing it through independent waste traders, farmers purchased street sweepings of horse droppings and the leavings of the city's roving pig population; manure from dairies and carriage stables; kitchen slop from hotels and restaurants; bones and offal from dead horses; carcass wastes from butchers and rendering operations; and household ash and organic castoffs.[14]

Delivering produce to the city, selling it, then in turn buying manure and offal represented what fertilizer historian Richard A. Wines refers to as "expanded recycling." For example, in the Northeast, as Wines explains, "Long Island and New York City in essence became a huge recycling system. The island provided energy for the city in the form of food, hay and fuel wood." Likewise the city returned nutrients for agriculture in the form of wastes.[15]

In the succeeding decades farmers grew reliant on new, more potent fertilizers like gypsum, guano and superphosphates, all manufactured substances. This crucial shift further disconnected farms from the reuse of wastes and signaled the beginning of true capitalist agriculture, distinct from earlier practices because productivity was now reliant on additional inputs that had to be purchased. As a result, castoffs were no longer as intensively recycled, marking another dramatic change in waste treatment. Now money, not shit, was the key. And growing competition made concentrated soil amendments mandatory for all; if farmers stuck with older practices of waste reuse they could not turn out the copious amounts of corn, potatoes, wheat and hay the market now demanded.[16]

Thus began the tendency toward disequilibrium in rural–urban relations, as observed by one nineteenth-century philosopher:

> Capitalist production collects the population together in great centres, and causes the urban population to achieve an ever-growing preponderance. This has two results. On the one hand it concentrates the historical motive power of society; on the other hand, it disturbs the metabolic interaction between man and the earth, i.e. it prevents the return to the soil of its constituent elements consumed by man in the form of food

> and clothing; hence it hinders the operation of the eternal natural condition of the lasting fertility of the soil.[17]

As farming shifted away from utilizing the swells of urban bilge, other parts of the new industrial economy moved in to absorb the mess. Nineteenth-century industry extensively utilized all kinds of garbage as feedstock, and supplying businesses with rubbish provided work for armies of scavengers.

Living and Laboring in Filth

The dirty ways to eke out a living in a nineteenth-century city were as numerous as the contents of a waste bin. Poor and immigrant children earned nickels from pedestrians in exchange for sweeping pathways through the accumulations of ash and grime blocking streets.[18] Cart men collected rags, bottles and rubber to sell to factories. And workers sifted through ash at dump sites to extract all variety of reusable and repairable items. Discarding, even into the latter part of the nineteenth century, was merely a system of recycling. In this era almost nothing went to waste.

The bounty of urban horse carcasses and abattoir wastes was the source of particularly useful commodities. Better specimens of bones were carved into handles and buttons, while smaller bits were ground and charred for use in sugar refining or in making commercial fertilizer. And almost all the salvaged marrow could be used for making soap and candles. Blood and offal were also highly valued. According to Benjamin Miller in *Fat of the Land*: "Sugar refiners and fertilizer manufacturers wanted blood. Flesh, boiled, was another material sought by the tallow industries. Hooves were used for gelatin, and in the dye known as Prussian blue. Hides and hair had their uses. Whatever remained was hog food."[19]

Likewise, cloth was used, reused and transformed until it almost disappeared. Worn sheets were "turned"—cut lengthwise down the middle, flipped, and restitched so that the thinning center was transferred to the less-used edges.[20] When too worn or stained, sheets and other household fabrics like tablecloths and curtains were reincarnated as pillowcases, bandages, diapers, sanitary napkins, and washrags. Finally, when these tired and threadbare swatches were beyond redemption, they were bartered or sold to cart men who traveled city streets or circuits of rural villages trading metal pots, cups, knives, and bolts of cloth in exchange for a range of discards including rags to sell to paper mills.[21]

In urban centers, cart men were often hired by businesses and individual householders to regularly collect wastes.These laborers deposited their haul at dumping lots or piers where the muck was processed. In towns near waterways, another type of refuse worker, called a "scow trimmer," loaded garbage onto barges for transportation to treatment plants or, quite often, for ocean dumping. Mostly, the trimmer's job entailed leveling out the rubbish on deck as cart men off-loaded from the docks above. Once at sea, scow trimmers were also charged with shoveling the heaps of ash and garbage off the vessel into the water. Along with their wage, these workers could typically keep whatever they culled, like fabric, which was passed along to "rag pickers" for sorting.[22]

One account sketched the macabre scene at a New York City loading pier: "Now the people who trim these scows at the various dumps are on the tops of those scows from morning till night. The ashes, garbage, and refuse is almost literally deposited upon the backs and heads of these people who stand there as fast as each cart load is dumped and spread. . . . These people also live, eat, sleep and have their sole habitation under the dumps of the Street Cleaning Department."[23] Indeed, many early New York City garbage workers, too poor to pay rent while they kept the city clean and created value from its filth, had to live in the space beneath city piers, the urban bowels, amidst the stinking piles of effluent.

Rag pickers and other scavengers also worked inside tenements. Streets like Bottle Alley and Ragpicker's Row in New York's Five Points neighborhood reflected these garbage-related labors carried out at home. *New York by Gaslight*, a chronicle of New York during the Industrial Revolution, offers this description: "A junkman's cellar in the front house opens widely to the street, and peering down, one may see a score of men and women half buried in dirty rags and paper, which they are gathering up and putting into bales for the paper mills. . . . [They are also] sorting old rags or cutting up old coats . . . that are too rotten to wear."[24]

The constant reuse of materials was not always carried out in the most ethical fashion. In 1863, Dr. Ezra Pulling, a volunteer sanitation inspector in New York City, described the heinous recycling of city wastes. Rotting food scraps, stale bread, and dead cats, rats, and puppies were "introduced into a *post mortem* fellowship" to produce sausage that was sold at sailors' boarding houses. Whatever dregs remained—according to Pulling, "a *debris* of material too thoroughly saturated with street-mire to be considered savory"—were sold to the makers of cheap coffee, who desiccated and partially carbonized it before mixing it with chicory.[25]

Under the Manhattan dumping piers at East 17th Street, according to one testimony, workers dined on half-eaten chicken legs and sausage ends sifted off a cart full of blood and pus-soaked bandages from the local hospital.[26] Apparently it was not bad enough for the poor to live and work in garbage; economic circumstance compelled them to eat it as well.

Class, Disease and Reform

During the nineteenth century's worst cholera and yellow fever epidemics, the death toll in New York City's slums—where half the city's residents lived—reached horrifying heights. One of the city's worst tenements recorded a fatality rate of 20 percent during a midcentury cholera onslaught.[27] The chief cause of these lethal outbreaks was constant exposure to filth. Scavengers and garbage workers were not the only ones forced to live in wretched conditions. *Hot Corn*, a popular nineteenth-century New York City paper, printed this grim account of a tenement: "Saturate your handkerchief with camphor, so that you can endure the horrid stench, and enter. Grope your way through the long, narrow passage—turn to the right, up the dark and dangerous stairs; be careful where you place your foot around the lower step, or in the corners of the broad stairs, for it is more than shoe-mouth deep of steaming filth."[28]

As political refugees and workers flocked to U.S. cities, urban living conditions seemed to grow more vile. Slumlords like New York's John Jacob Astor responded to the influx of immigrants by partitioning larger rooms and building upward, expanding existing one- and two-story apartment houses to accommodate ever more tenants. Atop these rickety structures landlords would add new stories of brick and mortar with little or no framing, heedless of engineering and safety concerns.[29] Borrowing the ideas of geographer Grey Brechin, one can think of these structures as mineshafts that run up, rather than down, extracting rent rather than ore.[30] These overcrowded, vertical shanties were veritable disease incubators thanks to insufficient light, ventilation, and sewage and garbage disposal.[31]

Odious and lethal, slums and the diseases they harbored attracted the concern of members of the growing bourgeoisie, who formed civic organizations that eventually spawned the Progressive urban reform movement. Groups founded in the mid-nineteenth century by wealthy activists, New York's Association for the Improvement of the Condition of the Poor (AICP) prominent among them, focused their attention on documenting and cleaning filthy urban tenements. Since invisible killers like cholera could easily swarm to upscale districts, slaying the gentle-born as well as the city's lumpen, these volunteer efforts prioritized cleaning the quarters of the city's poor.

Class was always an element in tidying up city streets. In 1849, at the behest of the offended, sanitation-minded well-to-do, and with the summer promising a particularly gruesome outbreak of cholera, New York City officials took on the hog problem.[32] Swine were a common feature of early industrial towns; workers still accustomed to raising their own food kept pigs that wandered freely, feeding off the piles of refuse tossed onto neighborhood lanes. For the better classes, street waste was a breeding ground for disease, a nuisance, and a symbol of social decay. Lawyers and merchants wanted hogs exiled, while the poor defended the right to graze their future chops and hams on the free bounty of garbage-strewn streets.

In the body of the pig, garbage became caught up in the politics of class struggle. Through hogs, street trash served as a type of urban commons, a means of survival for the poor, unjust as that might be. Despite anti-hog laws, urban swine herders persisted out of necessity and as a form of resistance to wage labor: garbage-fed hogs were a cheap source of protein that allowed the poor to be less reliant on laboring for wages. As law historian Hendrick Hartog notes, "A working class without its pigs would be that much more dependent on the market and employers, that much more controllable in situations of labor conflict."[33]

A few decades earlier, in 1821, New York's Black and Irish women in what is now the Lower East Side and Chinatown had mounted a successful defense of their pigs against similar crackdowns. Priggish anti-swine officials subsequently attempted to round up the bacon again in 1825, 1826, 1830 and 1832. The long war of the pig peaked in 1849 with the massive raid by club-wielding cops. Going house to house, they roused porkers from cellars and courtyards where their owners were hiding them, driving the swine to unsettled tracts of land uptown.[34] It took several more years of repression before a truly pig-free zone was established in Manhattan; beyond and in spite of this *cordon sanitaire*, the working classes persisted in keeping their edible beasts into the 1860s.[35]

Despite the 1849 hog sweep, a vicious cholera outbreak ensued, only the second ever to hit the United States. The pestilence thrived in the city's trash-strewn slums and crept into its better quarters, striking five thousand residents, who defecated and vomited to the point of fatal dehydration.[36] While the poor could be spatially contained, it was now clear that disease could not. An 1853 report by the New York Association for the Improvement of the Condition of the Poor advised that "it is a well established fact, that diseases are not confined to the localities where they originate, but widely diffuse their poisonous miasma. Hence, though the poor may fall in greater numbers because of their nearer proximity to the causes of disease . . . the rich, who inhabit the splendid squares and spacious streets . . . often become the victims of the same disorders which afflict their poorer brethren."[37]

Instead of tackling the economic polarization of the new industrial economy as the source of the problem, these civic groups approached the death and suffering of slum dwellers as a moral question: filth was connected to spiritual decay. "There is a most fatal and certain connection between physical uncleanliness and moral pollution," concluded one widely read journal at the time.[38] In mid-century an influential clergyman speaking to a group of civic activists stated that "where the body is unclean, and the dwelling wretched, there is commonly a correspondingly [sic] moral degradation."[39] Early reform groups reasoned that cleaning the offal and excrement from streets, alleys and tenements would set the situation right.

The Filth of War

The budding sanitation reform movement received an organizational push with the outbreak of the Civil War. As the war began, a private entity called the United States Sanitary Commission was formed by well-to-do volunteers to coordinate humanitarian logistics and sanitation in military camps. A literature and current affairs magazine, *North American Review,* explained the body's task as "the labor of calling the attention of the national army, by a system of inquiry and advice, to the peril of neglecting the conditions of health, and to the immense advantages of the strictest regard to sanitary and hygienic principles."[40] After all, almost as many Union soldiers fell to infection as were killed by the Confederacy.

In July 1863, sanitation got another, perhaps unlikely, boost in the form of the massive and violent New York City Draft Riots. That summer the streets were swamped with filth and the general mortality rate, much of it due to disease, was among the highest of the world's developed cities—one in thirty-six.[41] The sweltering, unrelenting heat with its concomitant quotient of pestilence and discomfort had everyone on edge. Names of the Union dead were posted daily in the papers for friends and relatives to study nervously. Then news arrived that all as yet unconscripted draftees would "have ten days to procure a substitute, pay three hundred dollars, or . . . take his place in the ranks of the Grand Army of the Republic."[42] In other words, the rich would get to escape the battle. It was the final straw. Thousands of enraged laborers, many of them Irish immigrants, exploded into the streets.

Along with horrific racial violence against African Americans—a reported eleven men were lynched while hundreds, including women and children, were forced from their homes and businesses—the mobs targeted their class enemies, those responsible for the draft and its enforcement.[43] Using brickbats, torches and guns, the rioters attacked draft offices, set them ablaze and destroyed their conscription ledgers. When troops arrived to suppress the mobs, rioters took over local armories, burning one to the ground; some distributed weapons and started killing soldiers. The mob beat to death an army colonel named O'Brien, dragged his corpse through the streets, then hung it from a lamppost.[44] They seized the provost-marshal's office, cut the building's telegraph lines, brutally bludgeoned the deputy provost-marshal and torched the building.[45] The rioters also sacked the mayor's residence.[46] And, after laying siege to the abolitionist *Tribune* offices, angry crowds viciously beat two different men whom they mistook for Horace Greeley, the newspaper's owner.[47]

Working simultaneously in dispersed groups across the city, the rioters numbered in the thousands; they destroyed factories, looted wealthy homes and upscale stores, assaulted gentlemen in fine carriages, and burned down mansions.[48] As the *New York Times* reported during the riot, "All vehemently protested against the '$300 clause' and were willing to be drafted, if the rich man would be made to shoulder the musket the same as they."[49]

With local police and soldiers flatly outnumbered, both the governor and the mayor beseeched Gotham's law-abiding citizens to form volunteer associations of "armed squads in their respective neighborhoods to protect their property and the peace of the City."[50] The carnage went on for four days until troops were called from the front lines and the riots were put down with volleys of lead.[51] In the end, an unknown number lay dead, including women and children. It remains one of the largest civil disturbances in U.S. history.[52]

From the Ashes

In the aftermath of the Draft Riots, civic-minded members of the middle and upper classes (perhaps those who had taken up arms to control the rebellion) began to connect cause and effect in new ways. In the eyes of forward-thinking elites, disease, garbage, crime, poverty and rebellion formed a coherent field of threat. The *Tribune* described the landscape this way:

> In those places garbage steams its poison in the sun; there thieves and prostitutes congregate and are made; there are besotted creatures who roll up blind masses of votes for the rulers who are a curse to us; there are the deaths that swell our mortality reports; from there come our enormous taxes in good part; there disease lurks, and there is the daily food of pestilence awaiting its coming.[53]

If conditions got too bad for the city's "dangerous classes," social order itself was threatened. Charles Loring Brace wrote fearfully of this "vast and ignorant multitude, who, in prosperous times, just keep their heads above water, who are pressed down by poverty or misfortune, and who look with envy and greed at the signs of wealth all around them."[54] The worse the misery and filth those at the bottom were subjected to, the greater the risk for volatile sickness and rebellion.

The reform efforts that followed in the wake of the riots centered on a renewed concern about sanitation. Directly and indirectly, refuse and public health were imbricated with the general project of shoring up social order. Just six months after the Draft Riots, a well-known reformer, Dr. John H. Griscom, joined some of the city's wealthiest lawyers, merchants and real estate tycoons, like John Jacob Astor Jr., Peter Cooper, August Belmont and Hamilton Fish, to form the powerful Citizens Association of New York "for purposes of public usefulness."[55] These reformers did have sound ideas: Twenty years prior, corrupt aldermen removed Griscom from his post as City Inspector because of his outrageous suggestions, like piping water from the Croton reservoir into every city home, enforcing building codes, and staffing health inspection jobs with medical experts rather than grafting political hacks.[56]

Groups like the Citizens Association also formed in response to rampant corruption in local government, which, among other catastrophes, created dysfunctional sanitation systems. The epic mounds of filth piled on nineteenth-century streets are notorious even today. Lacking any consistent policies on trash treatment, most local schemes were haphazard at best. At worst they were nonoperational, mired in backroom corruption thanks to dishonest politicians and the ineffectual contract system for waste removal. According to public health historian Charles E. Rosenberg, "The contracts were political manna and it was assumed that the contractor would make no more than token efforts to fulfil [sic] the duties which he had agreed to perform."[57] A particularly egregious mid-century contractor who was paid handsomely by New York City—then under Tammany control—to deliver barge loads of refuse to a Jamaica Bay processing plant, instead rarely left the pier and just dumped the muck through a trap door right into the river.[58]

Chicago was no different. A report from later in the century revealed serious negligence: "It is practically a universal fact that the quality of contract work is depreciated. Such efficiency as is obtained must be secured by the difficult and halting method of perpetual nagging by inspectors, with imposition of threats and penalties. How effective this method has been Chicago's residents too well know. As for the work of the past summer, when one begins to talk about its quality the subject simply becomes ludicrous. The contractor made his mileage, but the streets were not cleaned."[59]

Thanks to its dysfunctional contract system, Cincinnati's public lanes were so clogged that, at times, carriage owners left their vehicles at home and reverted to using streetcars. In response to these conditions, the city's most influential businessmen joined forces to protest the local Health Department, declaring, "Through a long period of municipal thievery and maladministration the streets of the city have gradually become more and more filthy. The Health Department has . . . become so corrupt as to be a public nuisance."[60]

Epidemic venality, legerdemain and bald-faced lying by official municipal health boards, departments and commissions meant that America's streets stayed dirty and disease-festered, and people died.

Addressing this problem in New York City, the Citizens Association's first major intervention was a massive investigation that sent a host of doctors to canvas local tenements, fetid shanties and illegal settlements under garbage dumping piers.[61] Although the procedures were based on "home visiting" by religious organizations and previous groups like the AICP, the Citizens Association study was a landmark in the history of epidemiology.[62] It also allowed the powerful a detailed view of the very population that had spearheaded the Draft Riots.

A year after the uprising, the Citizens Association filed a report that clearly linked trash and squalor with social upheaval. According to the inquiry: "The mobs that held fearful sway in our city during the memorable outbreak of violence in the month of July 1863 were gathered in the overcrowded and neglected quarters of the city. . . . 'The high brick blocks . . . seemed to be literally *hives of sickness and vice*. It was wonderful to see, and difficult to believe, that so much misery, disease, and wretchedness can be huddled together and hidden by high walls, unvisited and unthought [sic] of, so near our own abodes.'"[63] Its exposé of such brutal living conditions helped the Citizens Association win key sanitation reforms. In 1866, New York State passed the Metropolitan Health Bill, which provided comprehensive sanitation regulations and placed enforcement in the hands of state professionals. Similar laws were subsequently enacted around the country.[64]

The Birth of Industrial Garbage

Along with a wave of reform, the Civil War unleashed massive economic and technological transformations. The appetites of total warfare demanded industrial organization, consolidation and state intervention on an unprecedented scale. Weapons, train cars, uniforms, rations, boots, medicine, paper, coal and all manner of commodities were fed into the apocalyptic furnace. As with later wars, victory depended on production, and the new scale of battle-related industrialization helped to profoundly reshape American manufacturing, triggering unprecedented change in the quantity and quality of the country's garbage.

Commodities that had once been made in small workshops or at home were now produced on a mass scale. As the eminent historian of domesticity and waste Susan Strasser points out, in the United States there were over six hundred soap-making companies in 1857, each with an average of five employees. Cincinnati alone had twenty-five soap factories. After the Civil War, commercial soap production doubled but the number of firms plummeted; just three brands dominated the market: Colgate, Procter & Gamble, and Enoch Morgan's Sons.[65] And while the 1841 edition of Catherine Beecher's *Treatise on Domestic Economy* instructed readers on making soap and candles from household fats and tallow, by 1869 new editions of the book omitted these instructions, commenting that "formerly, in New England, soap and candles were to be made in each separate family: now, comparatively few take this toil upon them."[66]

Consolidation in manufacturing brought economies of scale and lower prices, which in turn lured growing numbers away from home production and into the general store.[67] This transformation was noted by one turn-of-the-century writer in the journal *The Chautauquan*: "Now the cabinet-maker having deserted his little shop has moved up to town, and become an employee in a great manufacturing establishment, and the housewife, having ceased entirely from producing, has learned to content herself with buying and using. The producer of household stuff today is neither housewife nor village cabinet-maker, but a factory 'hand.'"[68]

Industrialization also hastened the rate of urbanization. With the productive base shifting from agricultural to factory labor, workers from the countryside (be it from Italy or Pennsylvania) poured into

U.S. cities. Philadelphia grew fivefold between 1800 and 1850, while between 1820 and 1850 the populations in Boston and Baltimore quadrupled. New York City mushroomed from 60,000 in 1800 to almost a million sixty years later. By 1890, New York had grown by an additional 30 percent, and by 1900, with the incorporation of all five boroughs, the city was home to three and a half million souls.[69]

This escalating concentration of people in urban centers had profound effects on daily life and its wastes. As Dominique LaPorte explains in *History of Shit*, "The creation and acceleration of the division between town and country—a dichotomy that enfolds the fundamental head/tail reciprocity of shit and gold—is an effect of what is thus aptly known as primitive accumulation."[70] Not only were rural areas becoming less the domain of the subsistence farmer and more the site of intensified, capitalist food production, but the new ways of handling wastes signaled a similar transformation in the domestic arena. The treatment of excrement and other wastes by professionals, instead of by those who generated it, was a change that came as industrialization and the market system took greater hold in cities. Garbage as we know it today is one outcome of a fully realized capitalist system.

These new wage-earning city dwellers were more likely to purchase—rather than produce their own—milk, bread, clothing and other staples of daily life. The spatial and temporal characteristics of city living crucially shaped the nature of garbage. Due to long hours on the job, industrial laborers had less time for repairing and rendering what would otherwise be "waste." And crowded tenements left little space for storing scraps like fat, fabric and ash. Such changes meant more garbage. As industry developed, burdensome domestic chores eased, consumerism began, and so too did modern waste.

Getting Clean

Faced with the persistence of crooked politicians and colossal amounts of trash, scores of civic reform groups convened around the nation in the 1890s. Among them were the Women's Health Protective League of Brooklyn; the Civic Improvement League of St. Louis; the Neighborhood Union, organized among African American women in Atlanta; the Municipal Order League of Chicago; the Citizens Health Committee of San Francisco; the Woman's Club of Dayton, Ohio; the Philadelphia Municipal Association; and the Woman's Municipal League of New York.[71] Like their earlier counterparts, this new generation of do-gooders wanted to influence local governments and improve public health and sanitation, which they did through a variety of methods.

Some civic groups took it upon themselves to police the street-cleaning practices of contractors and municipalities. Jane Addams's Hull House Women's Club busted three of Chicago's city trash inspectors on over a thousand different violations in their neighborhood alone.[72] Others focused on influencing individual behavior, going into the homes, usually of the immigrant poor, to teach sanitation practices. One volunteer group explained its ethos: "Outside neatness, cleanliness and freshness . . . are the natural complement and completion of inside order."[73]

New York City's newest powerhouse civic organization, the Committee of Seventy was formed just after the jarring 1893 economic crash. One of the group's top priorities was the staggering trash problem, which they took on by intervening in the 1894 mayor's race. The Committee of Seventy campaigned hard against the Tammany candidate, and, in turn, the triumphant reform mayor brought in one of the most famous and visionary figures in garbage handling: Colonel George E. Waring, Jr.

Sanitation was tightly linked to public health and medicine, and, as it took on increasing autonomy, was still considered a scientific endeavor. As explained by plumbing historian Maureen Ogle, "The twin cultures of scientism and professionalism naturally spilled over into the field of sanitation and spawned the appearance of the sanitarians, a group of professionals dedicated to discovering and categorizing the principles of what they called sanitary science."[74] George E. Waring was one of the foremost sanitarians of his day.

Famous for gallivanting on his well-groomed steed, wearing a pith helmet, riding boots and handlebar moustache, the eccentric Civil War veteran had many talents. He bred the first widely grown table tomato, known as the "trophy," and designed the drainage system for Central Park under Fredrick Law Olmstead.[75] In 1895, "the Colonel," as Waring liked to be called, was made commissioner of the fledgling New York City Street-Cleaning Department. The agency was responsible for sweeping city lanes and collecting refuse from businesses and households. In his four short years running the city's war on filth, Waring actually got the place clean by ushering in the most innovative sanitation reforms the country had yet seen.[76]

To build morale and esprit de corps, the Colonel gave his public sanitation force a new paramilitary form and structure. Dubbed the

White Wings, Waring's workers wore starched white uniforms and pushed brand new, custom-designed (by his wife) collection cans. Their pay and numbers were also boosted. This two-thousand-strong sanitation army paraded through the streets on holidays and at public events, drilling in synchronized marches accompanied by a band. The public loved the White Wings; with their seemingly constant presence and reassuring medico-military look, they left behind safer, sanitized streets.[77]

The Colonel also brought discipline and a comprehensive discard sorting and reuse system to city waste handling. As explained by Waring, first, he barred independent gleaning and restricted city employees from "sorting over or picking refuse, or permitting others to do so."[78] This enclosure of the garbage commons was a crackdown on the poor who labored in waste, and meant that scavengers could no longer work the streets; gone in many areas was the individual cart hauler, "who jangles his string of bells through the streets." According to the Colonel, only licensed secondhand dealers, "whose transactions can be held under proper supervision," were allowed to gather refuse from private homes and businesses.[79]

Collection from dwellings, shops and offices came under strictures that were designed to keep wastes off the street and shut out unlicensed sorters. Waring's new system required New Yorkers to engage in "source separation," the segregation of ashes and food scraps from other refuse materials. And, once they were sorted, instead of putting these discards on the street for pick-up, residents placed an elegant "call card" discreetly in the front window.[80] Fully licensed handlers then retrieved the trash and hauled it to one of seven piers, where barges were loaded before heading to Barren Island in Jamaica Bay.

The tiny island, measuring a half-mile wide and one and a half miles long, was home to the New York Sanitary Utilization Company, a salvaging factory established under Waring's new garbage program. Barren Island also contained a village of eight hundred people—mostly poor African Americans, Irish, Italians, and Poles—who lived and worked at one of the island's five massive discard treatment facilities. Conditions on isolated Barren Island were tough, as residents were cut off from the outside world save for the weekly mail boat and a constant stream of trash-filled barges. The adults relied on four saloons for distraction, while the youngsters, who also worked in the various processing plants, had a one-room schoolhouse as their only diversion.[81]

Denizens of this village cum refinery cum dump spent their days combing through ash heaps for salvageables. At Waring's Utilization Company, sorting began on the island's beaches, which were covered with docks where scows off-loaded their dross onto a 104-foot conveyor belt. Workers sat along this refuse artery scouring the passing mounds of discards for recoverables. A journalist who saw the operation gave this description: "One picker selects manila paper, another shoes, another bottles, cans and metals another cloth and rags, until finally fully sixty per cent of the material which New York householders consider worthless is picked out as worth saving."[82]

Organic wastes went through the "Merz" process, more commonly called "reduction," which entailed cooking then compressing these castoffs to make grease and fertilizer. Pioneered on a smaller scope in Vienna, the Merz method was used on a mass scale only in the United States. As two leading sanitarians, Rudolph Hering and Samuel A. Greeley, proposed, "The greater wastefulness of the American people is one reason for this development, as it produces a garbage rich in recoverable elements."[83] After the country's first reduction plant opened in Buffalo in 1886, dozens more sprang up in Rochester, Toledo, Chicago, Los Angeles and elsewhere.[84]

Barren Island's reduction process sent rejectamenta like food slop, offal and dead animals into huge diabolical vats where it was boiled. The process was detailed in an 1897 issue of *Scientific American*:

> The cooking is allowed to go on for a period of eight to ten hours, until the garbage is thoroughly disintegrated and reduced to a pulplike consistency, all germs in the meantime being thoroughly destroyed. . . . This is then run into the presses, where it is subjected to a pressure of 250 tons. The presses work at a very slow speed, and it takes about three-quarters of an hour to compress the mass from 4 feet to 18 inches.[85]

The liquid extracted from the compression process was "used for the manufacture of soap, candles, etc., and is known commercially as 'soap grease'," while the solids, called "tankage," went into mixtures for fertilizer.[86] Only that which was utterly unsalvageable such as items broken beyond repair, or the bedding of those stricken with cholera, yellow fever or other lethal diseases, was sent to the island's incinerators. In one form or another, working with the city or around it, salvaging businesses on Barren Island operated steadily from the mid-nineteenth century until 1936, when Robert Moses, New York City's *über* planner, evicted all residents and shut the place down.[87]

Early conservationists embraced the reduction process because it helped "restore to the ground in some form of fertilization the constituent parts which have been taken from it in the shape of vegetable substances."[88] The deeper motivation for Waring was capturing resources; he considered disposal without culling reusables profligate, not for environmental reasons, but for economic ones. On incineration, the Colonel commented: "Cremation means the destruction and loss of matter which may be converted into a source of revenue."[89] One of his lieutenants, Macdonough Craven, went so far as to argue that "destruction of plant food should be prohibited by law."[90] A journal agreed: "It has lately been shown that there is sufficient commercial value in a considerable portion of the city refuse to more than pay for the cost of its collection."[91] These were evangelical modernizers; efficiency was godliness. The bottom line was central, and salvaging by municipal carters saved money.

Before Waring's institutionalized scavenging efforts, Americans commonly separated their castoffs, whether at the farm, informally in the urban household, or at the dump. Today, the in-home separation of wastes is often mistakenly considered a new routine, but sorting according to formal refuse taxonomy remained typical in the late nineteenth and early twentieth centuries.

In 1910, pioneer sanitarian William F. Morse wrote, "It is commonly the practice in American towns to make a separation in the household of the three classes of waste. . . . The householder is required to have three receptacles, for garbage, ashes and rubbish."[92] At the time "garbage" was often defined as "vegetable matter and table waste." The category of "ashes" could also include "floor sweepings, broken glass, discarded kitchen ware, tin cans and worn-out furniture." And "rubbish" usually referred to "paper, card-board boxes, rags, bottles, metals, old clothes, shoes and rubbers."[93] The kinds of commodities that constituted trash had changed drastically over the previous century, largely due to industrial production. New garbage handling methods that would follow on the trailblazing work of Colonel Waring now had to factor in the wide variety and growing quantity of discards.

As for the Colonel, his reign was short. Corruption and the heaps of stagnant dross that came with it soon returned. Ousted from his post in 1898 by the newly reinstated Tammany administration, Waring died a year later from yellow fever, which he had contracted while working in Cuba.[94] But his organizational and technical innovations lived on. Reform efforts culminated in a spate of rubbish-oriented municipal planning; in 1880, 24 percent of American cities offered some kind of refuse collection and disposal system, and by 1914 that number had doubled.[95]

New York's Fifth Street before (above) and after (below) the initiation of routine street cleaning, ca 1895. (Courtesy of the Museum of the City of New York, The Jacob A. Riis Collection)

3

Rationalized Waste

An epic theme is woven into the history of Sanitary Engineering. Something of romance has entered into its rapid progress and the story of its marvelous achievements in a period embraced by so short a span of years.

CHARLES GILMAN HYDE, 1934

The breakthrough science of bacteriology swept the sanitation field at the turn of the century, providing answers to a desperate populace weary of fear and death. An elaboration of germ theory, bacteriology was the outcome of decades of work by Louis Pasteur, Joseph Lister, Robert Koch and others. They successfully identified microorganisms, modes of transmission and means of sterilization to guard against contamination. The science of bacteriology brought the awareness that disease did not generate spontaneously in substances like garbage and that it could be warded off with targeted cleansing and vaccines. This meant that keeping trash contained was no longer the key to exorcising the demons of epidemic outbreaks.[1]

Since plagues could now be substantially contained, disease interfered far less with commerce. During outbreaks in the previous century, entire neighborhoods, business districts, and ships might be quarantined, paralyzing vital elements of trade. Also, to escape the scourge, huge numbers would flee the cities, forcing business to grind to a halt.[2] An 1895 issue of the Chautauqua Institute's journal ran an article entitled "The World's Debt to Sanitary Science," highlighting the economic benefits of bacteriology: "The knowledge gained within the present century of the nature of the causes of certain contagious or infectious diseases, and of the mode of their spread, has been of immense benefit to the world from a purely commercial point of view. The old systems of quarantine, with their unnecessary hinderances [sic] to traffic and travel, have been done away with in most civilized countries."[3]

Coinciding with this freer flow of commodities and bodies was greater industrial output, which relied heavily on transportation. As a result, keeping streets clear of trash took on a new urgency and significance. No longer so directly connected to controlling contagion, clean streets were now a prerequisite for a burgeoning economy reliant on fast-paced production and distribution. In order to run smoothly, the market demanded clean, functional roadways.

Influenced by bacteriology and shifting commercial needs, two significant divisions occurred in the sanitation world. First, public health departments split from waste collection, a lasting break that signaled a more benign view of trash.[4] In 1905, sanitarian Rudolph Hering said sweeping city lanes was "for the purposes of preventing nuisances and inconveniences, rather than as a health measure."[5]

The second major division came when many municipalities segregated street sweeping from garbage collection and disposal, tasks that until this point had comprised different branches of a single endeavor. Separated from the more expensive and logistically complex task of collection and disposal of trash from homes, offices and shops, street cleaning became easier and cheaper. And with the introduction of electricity, automobiles and motorized sweeping equipment, there was much less ash, horse dung and debris from daily life cluttering the way.[6]

As waste handling shifted from the world of disease and death, it was increasingly viewed as a technical problem. In this climate, the profession of sanitation engineering emerged. Furthering the labors of earlier sanitarians, these specialized civil engineers worked primarily for local governments, designing refuse treatment and disposal systems. Sanitation engineers took a highly rationalized approach to garbage, viewing discards less as a resource and more as a logistical problem. In the engineer's hands, escalating trash output was not a sign of imbalance or profligacy; rather, these ever-rising flows merely constituted a substance that needed to be put in its proper place. In creating ever-more effective depositories for refuse, the earliest sanitation engineers embarked on what has become an ongoing aim in the profession: disappearing garbage.

Street Cleaning Days

The separation of street cleaning from trash collection and disposal coincided with increasing demands on the city's network of streets by commerce, along with a growing desire among the upper classes for aesthetic order in public spaces. The Progressive Era's City Beautiful movement, an effort to make urban environments more attractive through planning, cleaning and policing measures, was yet another form of enclosure. Thoroughfares occupied by the working class and poor as social spaces and extensions of their homes, say, for sleeping on hot summer nights, did not fit into this vision of beauty.[7] According to a journal from the 1930s, "Street cleansing is thus more than a publicly performed operation to rid streets of the dust, filth, litter and debris. . . . It is a service to civic aesthetics and to social psychology. . . . Of what avail would be a grand spaciousness, a fine lay-out, delightful vistas, and beautiful houses, if the whole of their settings were always messed over with the wastes of urban life? Then, too, the clean street aids property values and a dirty one depreciates them."[8]

The city street was also an elemental component of the industrial infrastructure, included in what planning theorist Edmund Preteceille refers to as "urban use values." According to Richard E. Fogelsong in *Planning the Capitalist City*, "These public facilities, although collectively owned and produced by the state, are as important to the system of private production as are privately produced and appropriated inputs."[9] Sanitation engineers, local governments and business owners understood that clean streets meant ease of movement—ensuring that people arrived easily at the factory, office or market, all necessary preconditions for the economy to function.[10] As Henri Lefebvre notes, "The street became a network organized for and by consumption."[11] Thus, keeping thoroughfares free of garbage took on ever-greater urgency.

While the middle class and elites had previously resisted paying higher taxes to fund the care of public lanes and boulevards, they were now reaping the benefits of the street's more commercially defined spatiality. A new willingness to fund public services, coupled with reformed local governments, contributed to a growing expectation that municipalities should be responsible for city sanitation. As elected officials obliged, more towns channeled resources toward keeping dung, ashes and chicken bones from blocking the wheels of economic progress.[12]

According to the U.S. Department of Commerce, in 1909 cities with populations over 300,000 cleaned half of all their paved streets on average five times per week, impressive even by today's standards.[13] By 1917, more than 90 percent of American municipalities directly oversaw the ablution of their thoroughfares, scrapping the previous century's corrupt contract systems, which had been so unreliable and costly.[14]

Ever true to their race and class biases, city officials directed street washing crews most often to those passageways seen as vital for local businesses, while "streets in outlying areas and in working-class and immigrant neighborhoods had much lower priorities and received less frequent service—and no service at all if the streets were unpaved."[15] Articles in publications on local government openly discussed such policies. For example, the *Municipal Journal* noted that in Philadelphia "prior to 1915 street cleaning work covered only the paved streets."[16] As reported in *American City*, Houston's teams worked on "boulevards and main thoroughfares daily" and made collections from the city's business district "daily . . . including Sunday."[17] Street tidiness, and by extension neighborhood sanitation, hinged on local leaders' allocation of resources, which frequently left the poor, working class and immigrants to live with a disproportionate amount of waste.

With the spread of electric streetcars and automobiles, the use of horses dramatically declined, profoundly simplifying the deterging of city lanes. At the turn of the century, U.S. urban streets were home to about three and a half million horses, which each produced on average twenty pounds of manure and two gallons of urine for every eight-hour workday.[18] So the shift to automobiles meant significantly less waste in public places. Formerly jammed with 80,000 horses and their leavings, Chicago saw a 300 percent increase in motorcars between 1910 and 1913.[19] There were almost 200,000 vehicles on Ohio's roads in 1915, and within two decades that number rose to more than 2 million.[20]

The increase in the number of automobiles made urban streets easier to keep clean and freer of stench, but also brought unforeseen environmental consequences: polution from engines and dependence on fossil fuels. As garbage historian Martin V. Melosi comments, "Little did they realize what the transition from animal power to mechanical power signaled for the city's physical environment. Trading manure and horse carcasses for hydrocarbons, noxious fumes, and waste heat from internal-combustion engines was no bargain."[21] Nevertheless, with reduced animal populations and more attention from municipal street sweeping departments, cities across the country were able to make progress in cleaning their filthy thoroughfares.

The Change at Home

The technological transformations in the decades after the turn of the century affected household waste as well. In 1910 only 10 percent of American dwellings had electricity, but by the end of the 1920s the majority of urban homes were wired.[22] Electric stoves and gas heating generated far less ash, which, as of 1918, comprised up to 75 percent of per capita wastes.[23] However, even as ash discharges were dwindling, the American household's overall trash output was escalating. New types of refuse like packaging, disposable paper products and discarded ready-made items joined familiar categories of wastes like food scraps in this growing reverse cornucopia.

Between 1903 and 1907 Pittsburgh's organic waste output almost doubled, while Milwaukee and Washington, D.C., dumped an additional 24 percent, and Cincinnati's garbage production surged by 31 percent.[24] During the 1920s the discards sent to one Rochester, New York, treatment plant increased by 200 percent.[25] By the early twentieth century, the urban working classes in America were tossing out more than twice as much garbage as their European counterparts.[26] As urban populations continued to increase—by 1920 more than half of the U.S. population lived in cities—so too did their waste.[27]

After the turn of the century, the nature of what people threw away also changed in significant ways as manufacturers began marketing mass-produced disposable goods for the first time. The paper industry conjured disposable products to sell as alternatives for already existing or more durable items: new rolls of toilet paper as a substitute for waste paper like the pages of the Sears, Roebuck catalog; paper cups (the now ubiquitous Dixie cup) in public drinking basins displaced shared cups; and paper straws supplanted the product's organic namesake, straw.[28] The still-new disposable sanitary napkin was marketed as an improvement over reusable rags and birdseye fabric.[29] There's no question many of these products brought greater comfort and cleanliness to their users, but they also brought more trash.

During this time, new forms of packaging also emerged. As manufacturers grew bigger and more centralized, they relied increasingly on the innovation of product packaging to keep their commodities clean and fresh. At the turn of the century, one of the earliest packaged items was National Biscuit's Uneeda Biscuit, sold in a patented carton of cardboard and waxed paper. A toothbrush manufacturer of the same era sold its product in a "sterilized" box so customers would not have to "buy from a fingered pile of dusty, germ-laden tooth brushes, handled by nobody knows who."[30]

In the early days of mass marketing, selling in bulk from barrels at small general stores could expose products to dirt and germs or result in the underweighing of items.[31] Demonstrating reliability by keeping products clean and undamaged helped earn the trust of consumers suspicious of goods made by producers they did not personally know. And premeasured packages suited large-scale industry's demands for accuracy and consistency because such a system allowed manufacturers, rather than consumers, to determine the amount of each purchased unit. The practice of shoppers filling their own pails or bottles with milk or beer from a tap or tank was on the wane as packaging led an industry-driven transformation to disposability.

Methods of Disposal

Before today's high-tech waste-handling processes came into being, refuse disposal was a gruesome affair, whether it be by land, sea or fire. Around the turn of the century, dead commodities most typically met their end at an "open pit" land dump. Located on the outskirts of towns, these waste depositories were set up on flat tracts, in gullies or ravines, on unused swampland or along coastlines. Just after the turn of the century, a survey of over one hundred cities revealed that most dumped ash, organic and other wastes on land.[32] Not surprisingly, the open pit had its drawbacks; it led to water contamination, toxic gases, spontaneous fires, and vermin infestation—ranging from flies and cockroaches to rats and bears.[33] At the end of the nineteenth century, one well-to-do citizen lodged this complaint regarding the hillocks of putrescence at the Rikers Island dump:

> It is a fact, as Judge Hall has borne witness here, that we had to get up in the night time, aroused from our sleep by the stench that filled our nostrils beyond endurance. . . . The silver on the door knobs, and other metallic substances in and about the houses in that section were turned and spotted with some black substance that would not respond to the polishing of servants or anybody else.
>
> Now if such things, such microbes, are floating in the air to be breathed into the lungs of humanity, must there not be some permanent injurious result? If metal can be so affected, what must the effect be upon a healthy lung?[34]

Another common method of early disposal was waterway dumping. Expedient for communities located on the coast, the Mississippi River, and the Great Lakes, this practice was horrendously destructive.[35] In the 1890s, when Seattle's residents complained about the nuisance caused by poor trash management, city officials began offloading garbage and ash into Puget Sound.[36] Around the same time, Chicago started dumping city discards and offal from its abattoirs three miles out in Lake Michigan. Even though the lake served as the city's water supply and commercial fishery, the local Department of Public Works deemed the solution both "convenient and suitable."[37]

At the turn of the century, the Louisiana Board of Health justified its water disposal method as follows:

> To dump the garbage of a large city into a running stream from which is also derived the water supply of the city, might seem, at first glance, a rather crude and imperfect, as well as unsanitary, method of getting rid of the city's waste; but when it is remembered that the Mississippi River is at this point a half mile wide, from fifty to one hundred feet deep, with an average current of three miles per hour . . . we may readily imagine how little influence a boat-load or two of garbage per day can have upon such an immense body of water in constant motion.[38]

In many seaside boroughs, the method worked well as long as the tide behaved. But if the waters turned, tons of clotted, rotting flotsam would foul beaches, contaminate drinking sources, and sit for days, bobbing and stewing around city piers. One report explained that the residents of Milwaukee, which also dumped into Lake Michigan, found particles of garbage in their drinking water. Nonetheless, towns like Boston, Cleveland and St. Louis all used water dumping.[39]

The sea and its tributaries were not, in fact, limitless, and the filth, shit and dross continually hurled into them soon caused major environmental, health and even economic problems. As early as 1888, ocean dumping was outlawed by the federal Marine Protection Act, although the ordinance was frequently ignored.[40] When marine disposal regulations were observed, effectively keeping offal away from inhabited shorelines, other safety issues could crop up. Water dumping was briefly abandoned in Oakland, California, after a ship and crew of nine men were lost at sea while hauling a load twenty miles past the Golden Gate, as was then required by law.[41]

Responding to the dangers of sea and river disgorging, U.S. Army Lieutenant H.J. Reilly came up with a solution, or rather, a new problem. In 1885, Reilly constructed the country's first incinerator on New York City's Governors Island. The technology was imported from England, where the contraptions were referred to as "destructors." Though highly problematic—the toxic stench of burning waste was nasty stuff and there was still plenty of ash to be disposed of afterward—this waste disposal method spread throughout the states. As one enthusiastic sanitarian exclaimed, "At last we have secured a means of destroying these substances [waste] and their ability to do evil."[42] By 1887, local officials had installed trash-ingesting furnaces at Wheeling, West Virginia; Allegheny, Pennsylvania; and Des Moines, Iowa. By the turn of the century, brand-new garbage "crematories" were belching smoke in sixty-five towns across the country.[43]

While incineration was popular, due to its inefficiencies it did not fully displace older methods of salvaging, hog feeding and dumping. Unlike today's highly flammable paper- and plastic-rich waste stream, garbage of old did not burn well because it was much wetter and denser. And since the facilities were so capital-intensive with their complex equipment and need for skilled labor, incineration was often bypassed for cheaper, more reliable disposal methods.

Enter the Engineer

Arising from turn-of-the-century reform activism, a new kind of specialist hit the streets—the sanitation engineer. The first "well organized, balanced curriculum in Sanitary Engineering" was offered at the Massachusetts Institute of Technology in 1894.[44] Influenced by the science of bacteriology, reformers shifted from targeting "poverty and behavioral degeneracy as the two primary threats to the public's health" to treating urban filth as a technical problem for skilled managers to solve.[45] The new field of sanitation engineering filled the bill nicely.

From early on, the engineering establishment deployed (as it still does) the motif of the engineer as an entirely rational expert seeking only what's right and best. In the 1930s, the trade journal *Municipal Sanitation* offered this encomium:

> The demands of his profession are such that the engineer must be sound in mind and body. He must have the moral strength to seek and to hold to the truth. "And ye shall know the truth, and the truth shall make you free." Only by knowing the truth can the engineer become free to determine the correct and untrammeled solutions of his intricate problems.[46]

Characterizing themselves as neutral and above politics immunized engineers against public criticism.

However, these technocrats were far from neutral. The engineering profession, of which sanitation is but one subset, emerged to fulfill the demands of capital and evolved with business and industry as its most influential forces. Engineering skills and the uses for this specific field of knowledge were only cost-effective and necessary for big budget projects like bridge, canal and railroad construction, which were the exclusive domain of industry and government.[47] Not surprisingly, powerful industrialists and businessmen held unrivaled sway over the field of engineering, and through this influence, they imbued garbage handling with their worldview on waste.

According to engineering historian Edwin T. Layton, business interests helped shape the culture of the discipline, largely through the conduit of professional associations. These powerbrokers have historically failed "to punish those who act against the public interest." Instead, the profession has discouraged practitioners from prioritizing the public good and moral integrity over free-market economic development.[48] The responsible use of natural resources, whether in the realm of manufacturing or trash disposal, often ran counter to capital's agenda of maximizing profits, so was frequently sidelined for a more industry-friendly approach.

In part, the anticommunism of the interwar years motivated this ethos. As Layton has observed, industry and business leaders who headed the professional engineering organizations "feared that the forces of bolshevism, class-consciousness, and disloyalty, if not arrested, would destroy the nation's social and political ideals."[49] And such political changes would undoubtedly impact revenues. Thus, guided by the hand of big capital, American engineering was recruited for the ideological crusade to reassert market primacy and laissez-faire government. This not only affected dam and road building but, of course, garbage treatment. So, despite potential or episodic turns to more sustainable systems, engineering's conceptualization and treatment of refuse increasingly reflected the interests and expectations of big business.

Politically Sanitized Wasting

J.C. Dawes, noted sanitation engineer and host of the first International Conference on Public Cleansing (in London in 1931), proposed at that gathering: "The first and greatest step is that the Local Authorities of all countries should formally recognize public cleansing work as being definitely technical in character and elevate it to the status of a recognized technical service."[50]

During the first three decades of the twentieth century, sanitation engineering brushed aside the notion that waste services should pay for themselves through salvaged materials. They also distanced themselves from the practice of separating organic discards to put back into the soil for conservation purposes. These modern trash technicians treated castoffs as an entirely different category of material that no longer had use-value. In the hands of the sanitation engineer, garbage was handled as an innocuous and unavoidable class of debris. Engineers also helped cultivate and improve spaces where annihilating discarded goods made sense, forging a new phase in disposal technologies.[51]

Sanitation engineers rarely challenged the fundamentals of a market system that pathologically wasted resources. Their acquiescence helped the American public accept growing quantities of garbage without contemplating its implications. Changing practices in the home reinforced this position; people were consuming more and throwing out escalating amounts while growing increasingly accustomed to having it whisked away by professionals. At this formative time, flush toilets, indoor plumbing, more consistent street cleaning and improved refuse disposal were all making the act of wasting easier both logistically and aesthetically.

It should be said that America's intensifying garbage output was fueled not only by these structural shifts, but also by a public that took the opportunity to use and waste in greater volumes. In part, the burgeoning disposable lifestyle made sense—after all, so much was made easier by new manufactured and throwaway items, especially for women, who engaged in the bulk of domestic work. As *The Chautauquan* pointed out earlier in the century, "The industrial changes . . . have not, of course, been without their compensating advantages. Of these the chief perhaps has come to the housewife, and consists in the opportunity to buy, ready made and at low cost, most of the articles which it used to be necessary for her to make at home."[52] The leisure time created by industrially produced goods and disposable items like sanitary napkins was no doubt ecstatically welcomed.

Another outcome of new and evolving sanitation systems was that households more frequently consumed and discarded behind closed doors, making individual families' acts of wasting private and discreet. This isolation fit neatly with the engineered disposal of refuse, and the growing cultural disconnect between trash and the larger system of industrial production. The result was an overall sanitization of garbage.

As Dominique Laporte has pointed out, the privy made the act of wasting private, while the creation of municipal infrastructures cast cleaning as public. The upshot of this was that wasting—whether bodily excretions, the flow of castoffs from a household, or the effluent of industry—was hidden from view and public consciousness. One's castoffs belonged to oneself; one was responsible for what one discarded, but where it came from and where it ended up became abstract and obscure. Through this veiling, wasting and waste were politically cleansed, thereby cleansing the system that produced so much waste. Laporte writes:

> The *privé*, disgusting place where one's little business is stealthily carried out while one rubs one's hands, becomes literally the place of primitive accumulation. It is the home of that small heap of shit which the subject tends to, maintains, even cherishes. The State, on the other hand, is the Grand Collector, the tax guzzler, the *cloaca maxima* that reigns over all that shit, channeling and purifying it, delegating a special corporation to collect it, hiding its places of business from sight. The State devises severe fines for proprietors who transgress laws ordering them to settle their affairs behind closed doors—those who, by letting the shit fly out their windows onto the street, might confirm the suspicion that "all this does not smell very good."[53]

Had they been obliged to deal with their castoffs more directly, perhaps greater numbers of people would have diagnosed the emerging system of mass production and consumption as deeply flawed. Not long after the turn of the century, a small group of dissenters did just that. Documented by historian Dolores Hayden, social experiments in low-waste living by feminist writers, architects and planners like Helen Campbell, Charlotte Perkins Gilman and Alice Constance Austin demonstrate the path not taken. Their more sustainable vision took the form of community kitchens and laundries, cooperative housekeeping, and not-for-profit cooked food services, all to pool resources and eliminate waste.[54] As Hayden explains, these women understood that sharing facilities would reduce consumption and discarding, thereby liberating them from the drudgery of maintaining the atomized, "isolated home" and freeing family incomes by lowering overall costs.[55]

Charlotte Perkins Gilman's 1903 book *The Home* incisively critiqued American industry's vision of private, single-family dwellings as "bloated buildings stuffed with a thousand superfluities."[56] Gilman explained "the waste of the endless repetition of the 'plant'" in a visionary way: "We pay rent for twenty kitchens where one kitchen would do. . . . We have to pay severally for all these stoves and dishes, tools and utensils, which, if properly supplied in one proper place instead of twenty, would cost far less to begin with. . . . There could hardly be devised a more wasteful way of doing necessary work than this domestic way."[57] In 1918, the *Ladies Home Journal* predicted: "It will not be long before . . . the cooperative kitchen has become a reality" in cities across the country.[58] And two years later the *Ladies Home Journal* concluded that "the private kitchen must go the way of the spinning wheel, of which it is the contemporary."[59]

After the Red Scare of 1919–1920, such ideas were rejected as communist and morally suspicious. So potent was the ensuing backlash that such moderate groups as the Young Women's Christian Association, the American Home Economics Association, and the League of Women Voters were targeted as deviant in the infamous "spider web" chart published by the War Department. The document listed these organizations and many others as part of a "red web" that threatened to obliterate the country by promoting collectivism and egalitarianism. Captains of industry pushed this line as well. Henry Ford reproduced the spider web chart in his daily newspaper, the *Dearborn Independent*, while the head of the National Association of Manufacturers intoned that mainstream American women's organizations had fallen under the influence of the infamous Soviet feminist Alexandra Kollontai.[60] The economic boom of the twenties dealt a final blow to these valiant experiments.[61]

Instead, mass production and consumption conquered America, facilitated by sanitation engineering. Greater efficiencies in trash collection and disposal led to more waste, not less. Consider that during World War I overall discard retrieval in U.S. cities declined by about 10 percent due to cuts in local services and wartime belt-tightening. Simultaneously and in contrast, the abundantly funded U.S. armed forces were churning out huge amounts of waste. The production of discards in army camps was 50 percent higher than the average urban household output.[62] In this instance, not only did refuse-making rise and fall with the health of the economy, it also escalated with the ease of organized collection and disposal.

Whether fully aware or not, engineers who took the mainstream approach of accepting trash as a natural, inevitable substance simply in need of safe treatment created structures that led to greater amounts of garbage. From this perspective, sanitary engineers, those professionals who formed the foundation of organized waste management in industrialized society, were by no means engaged in a neutral or apolitical practice.

The changes in the opening years of twentieth-century sanitation and garbage handling helped constitute an infrastructure crucial for mass consumption and discarding. The rise of the sanitation engineer was the capstone on a series of developments that rationalized and eased wasting, rendering it invisible. The science of bacteriology led to the reconceptualization of trash as an aesthetic and technical problem; new street cleaning methods prioritized the flows of commerce; and emerging garbage collection and disposal methods met the manufacturer's need to have mounting levels of disposable commodities, including packaging, tidily hauled away. Sanitation engineers together with emerging municipal refuse handling systems laid the groundwork for the coming golden era of garbage.

Household garbage being buried at the first modern sanitary landfill in the United States, Fresno, California, 1935. (Collection of James Martin)

4

Technological Fix: The Sanitary Landfill

The filthy cities of history, which sat in a clean countryside, are succeeded by clean cities encircled at some distance by their wastes.

KEVIN LYNCH, *Wasting Away*

Perhaps the most significant technological innovation in managing American trash was the development of the "sanitary landfill." Arising from a landscape of muck and swill, the earthen grave of the future was in its nascent stages just before World War II. While street cleaning was on its way to the ineffectual sweeping trucks and citation-laden alternate-side parking of today, efficient trash management did not come about until the 1930s. But during that decade, all U.S. cities with populations of more than 100,000 adopted some form of organized refuse collection and disposal.[1] Where to put the messy stuff was another question altogether.

With the Depression squeezing local budgets, cities switched from using expensive and complex treatment systems like reduction to various forms of land dumping, the cheapest method available.[2]

Alternative processes like grinding garbage, municipal hog feeding, and composting also lost out to the vastly less expensive landfill, as did incineration.

In the pits of gullies, ravines, swampy tracts and other low-lying areas, civil engineers used new earth-moving equipment and slim labor forces to create seemingly inexhaustible, low-cost depositories for society's discards. After World War II, many of the engineers who were trained in building sanitary landfills by the military returned stateside to take jobs in municipal sanitation departments. Through these channels the now ubiquitous sanitary landfill spread its infectious new gospel of "out of sight, out of mind" across the land.

Burned Out

Although incineration remained expensive, the trash burning industry peaked during the Depression, when there were incinerators up and running in more than six hundred cities around the country. Even though the number of incinerators would wane in the following decade, in the 1930s, refuse cremation was big business, thanks in part to New Deal funding and labor subsidies.[3]

As in the past, waste industry insiders extolled the virtues of burning. The trade press and the ever-growing professional sanitation associations vigorously promoted cremation as superior to land dumping. One journal praised it as "the most sanitary and efficient method of disposal."[4] In a manner that sounds familiar today, public protests were dismissed by the industry, as in this article from 1938: "The year has witnessed a few typical examples of rebellion by individuals or groups against the location of incinerators in their immediate vicinity. These objections arise from a complete misapprehension of the operating conditions of a modern high temperature plant. . . . "[5] Although unsuccessful in the 1930s, public opposition would play an important role in the fate of incineration later in the century.

Some producers of refuse-burning facilities sold their equipment with the promise of generating electricity and steam power from combustion, which would in turn create revenues; this method of burning was (and still is) called "resource recovery." The trade journal *Municipal Sanitation* enthused about the prospects, and notably revealed shifting attitudes toward discards in a 1938 article: "The ability of the incinerator to produce useful heat and power from a *useless* waste adds dramatic force to pleas for construction of such disposal facilities in municipalities."[6] But power generation from refuse never gained the momentum that it did in Europe, due in large part to abundant and low-cost energy supplies in the United States.[7]

Ultimately, incineration could not compete with the vastly less expensive method of land dumping. Even though lower labor costs were predicted for running incineration plants over reduction operations, landfills required even fewer and less skilled workers, a bottom line that was hard to beat and attractive to cash-strapped cities.[8] In the 1930s, it cost New York six cents per cubic yard to send its wastes to Rikers Island while the same quantity sent to an incinerator cost substantially more than twenty-three cents. In 1938, Portland paid double the price of burial to burn its castoffs.[9] Similarly, other methods successful before the 1930s would retreat in the face of low-cost landfill disposal.

The End of Other Means

In the early decades of the twentieth century, there were many alternatives to burning and burying. Dangerous and dreaded water dumping persisted, but so too did reduction and feeding organic wastes to hogs. Other innovations like municipal composting and grinding garbage were used to process wastes into soil amendments. But, like incineration, none of these methods could under-bid the landfill.

Water disgorging, that temptingly cheap and easy means of disposal, was coming under scrutiny yet again. Those who lived in beach communities and downstream from river dumping spots started complaining about the mucking up of their shores. According to a 1914 engineering report made to the Sanitary District of Chicago, residents along the Des Plaines River in Illinois disliked the "general nuisance, masses of dying or dead fish, and the general unfitness of the stream. . . . At times conditions have been almost intolerable for the people residing immediately along the bank of the river." Such fetid nearby waters threatened to bring down property values, angering vote-casting landowners.[10]

In 1933, ocean dumping of municipal wastes was again ruled illegal, this time by the U.S. Supreme Court in a case against New York's unabated water disposal. (Commercial and industrial wastes were exempted from the ban.)[11] This law did not necessarily stem from concern for environmental impacts like the health of aqua-

ecosystems. Rather, as had been noted in earlier years by the famous sanitarians Rudolph Hering and Samuel Greeley, the "fouling of beaches creates a nuisance that the public should not be asked to tolerate."[12]

Previously regarded as a cost-effective and sanitary form of waste treatment, reduction lost ground in the 1930s. *Civil Engineering* reported the plummeting price of garbage grease as a major factor in the 1938 closing of Detroit's reduction plant. When resale revenues fell, these facilities were forced to shut their doors. Reduction facilities in Boston; Indianapolis; Reading, Pennsylvania; Rochester and Syracuse, New York; and Royal Oak, Michigan, held out for a few more years, with Philadelphia's plant the last to close in 1959.[13]

Feeding waste to swine, despite the previous century's urban crackdowns, had not died out. Instead, the practice became institutionalized as places like Los Angeles; Kansas City; Worcester, Massachusetts; Sioux City, Iowa; Geneva, New York; and others separated their organic wastes to either send to municipal hog plantations or sell to private farmers. As reported in *Municipal Sanitation*, during World War I the U.S. Food and Drug Administration actually recommended giving garbage to pigs "as a food producing and waste conservation measure."[14] In the 1930s, hog feeding became the country's most prevalent method of food waste disposal.[15] And, by 1939, according to the U.S. Public Health Service, 52 percent of cities surveyed nationwide were "feeding garbage to swine."[16]

The much-touted facility in Worcester—one of sixty-one towns in Massachusetts that practiced some form of feeding wastes to hogs—was, according to *Municipal Sanitation*, "an outstanding example of the successful operation of a piggery by a municipality."[17] Southern California's Fontana Farms, a privately run concern known as "The World's Largest Hog Farm," fed its 46,000 porkers a daily diet of 400 to 600 tons of garbage from Los Angeles.[18]

In the 1930s, feeding swill to swine was attacked after the U.S. Public Health Service began to link trichinosis outbreaks in humans with garbage-fed hogs. Scientific studies at the time revealed raw-garbage feed to be a significant factor in the infection of hogs with the parasitic nematode that causes trichinosis. They also found that trichinosis could be transmitted to humans who ate undercooked meat from the infected pigs. Later, during World War II, the Public Health Service restricted interstate transport of raw garbage in an attempt to limit trichinosis outbreaks; this move surely impaired the marketing of organic wastes to hog farmers.

Then in 1953–55 sentiment against feeding garbage to hogs escalated again as an outbreak of the swine disease vesicular exanthema led to the slaughter of more than 400,000 of the country's porkers. In response, the Public Health Service and many state health departments banned outright the feeding of raw garbage to hogs. While cooking garbage would have overcome the problem, it was too expensive. Consequently, feeding slop to pigs steadily dropped off in the coming decade and by the early 1970s only about 4 percent of collected food waste was used as hog food.[19]

Stemming from an ongoing recognition of the value locked inside organic discards, experimental methods of municipal disposal like composting and grinding garbage were also used in the first half of the twentieth century. The idea was to divert organic wastes from incinerators and landfills, instead processing these castoffs into fertilizer. In the 1930s, *Municipal Sanitation* reported that the "progressive industrial city" of Lansing, Michigan, had opened a million-dollar joint sewage-and-garbage treatment plant after closing the city hog farm. The new low concrete-and-glass brick plant accepted municipal organic refuse, which was chopped by two hammermill grinders, then ejected directly through cast iron pipelines to "sludge digesters," where it was processed into soil amendment.[20] Around the same time, Baltimore also opened a city stool-and-slop plant.[21] Again, this method cost more than burying garbage and so never gained a substantial share of the disposal market.

Municipal composting was another alternative treatment method that fell by the wayside in the years before World War II. Large-scale composting was developed in Italy by Dr. Giuseppi Beccari and used in cities across that country beginning in 1914: the system was known as "fermentation" or "the Beccari method." This means of composting debuted in the United States in 1923 in Scarsdale, New York. The process was thus described: "Raw garbage is placed in a cubical shaped cell on a series of trays or gratings about two feet in depth. . . . After loading the cell for a few days, the cell is closed air tight and the anaerobic bacterial [sic] begin their work of breaking down the mass, liquefying the readily decomposable elements." In just over a month the organic wastes were transformed into humus, which was then dried, ground and "bagged and sold for fertilizing purposes."[22]

Beccari composting facilities were constructed in three more municipalities in the following years, and the method, although never widespread in the United States, persisted into the 1950s when Altoona, Pennsylvania, opened a successful plant.[23] But, like other alternative capital-intensive methods, municipal composting dwindled when forced to compete on a cost basis with garbage burial. As *Municipal Journal and Engineer* had foreseen years earlier: "It is possible to deposit garbage and refuse mixed, or even garbage alone if properly treated, on low land without creating a nuisance."[24] By the 1930s, land dumping was poised for massive expansion.[25]

The Sanitary Landfill

Because of the economies land disposal offered, there was incentive for garbage professionals to innovate cleaner and safer methods of burying wastes. While "open pit" dumps persisted, in the early decades of the twentieth century sanitary engineers honed more refined techniques like "plowing garbage into the soil" and "burial," primarily for organic discards. In applying these various methods, practitioners aimed to do more than simply take waste away from home, office and street; decomposition was of utmost importance. As the sanitarians Hering and Greeley explained: "When garbage is once properly buried on a tract of land, there should be no further offense from it. It will be gradually reduced by the slow processes of inodorous decay into suitable food for plants, and should leave the land much more valuable for agricultural purposes than before the garbage was applied."[26] However, breaking with this conception of disposal, sanitation engineers soon formulated filling practices that entombed all categories of waste together, creating the direct precursor to the sanitary landfill.[27]

The other major influence on the American sanitary landfill was the "controlled tip." This forebear was developed in Britain just after World War I as an alternative to expensive incineration. According to the trade publication *Engineering News-Record*, the method entailed depositing discards in "shallow layers, not more than 6 ft. in thickness" and forming "narrow strips called 'fingers,' which become in effect, soil-lined cells in which biological changes proceed independently."[28] A variation on this careful burial process would become the earliest version of the sanitary landfill now ubiquitous in the United States.

The country's first sanitary landfill was constructed in 1934 on the sparsely populated outskirts of Fresno, in California's San Joaquin Valley. The project was the brainchild of the city's commissioner of public works, an engineer named Jean Vincenz. During his ten-year service, Vincenz remade Fresno's infrastructure, building an airport and new city hall and improving the municipal sewage system. And, with no prior waste-handling experience, Vincenz opened Fresno's state-of-the-art sanitary landfill, a synthesis of older methods that would contribute significantly toward remaking American wasting.[29] A newspaper report from the time described the operation:

> The site selected for the sanitary fill is on city property near the sewage farm about 7 miles outside the city limits, where the average haul for the garbage trucks is about 9 miles. A full-revolving dragline stationed at the fill has a 40-ft. boom and a ½-cu.yd. bucket. This machine performs a three-fold function by (1) excavating a trench 20 ft. deep into which the garbage is dumped, (2) unloading the trucks into that trench, and then (3) covering over the dumped garbage with a layer of earth 1 to 2 ft. thick.[30]

The compaction by the dragline and dirt covering kept rodents out—the still new "vector theory" recognized rats as one of several culprits in disease transmission—and it kept the smell to a minimum.[31] Once the pile of tightly packed waste rose five feet above the original ground height it was capped with an additional twenty-four inches of earth as the next trench was dug. And just one worker, the dragline operator, could carry out most of the labor.

While streamlining the labor process was a major aspect of the sanitary landfill, it also permeated collection operations as well. As municipalities updated refuse retrieval methods, some began reassessing the *way* they collected, looking for faster methods to reduce costs even further. Municipalities like Fresno shifted from horse and mule carts to tractors and motorized trucks for hauling their trash.[32] Labor practices were also revolutionized, influenced by Fredrick Taylor's ideas on scientific management and the technical expertise of engineers.

Again, Fresno was at the fore. According to the trade journal *Engineering News-Record*, Vincenz and his Public Works Department applied Taylorist methods to investigate and fine-tune waste handling. Workers' collection routes were redesigned "to an area just large enough so that continual and conscientious effort would be necessary."[33] That way no work time was wasted. Vincenz also introduced new "low-body" trucks. As reported in the *Engineering News-Record*, these vehicles radically changed collection from the labor-intensive "common type of garbage truck in which as many as five steps must be taken by the collector to dump even his first load," to a simple one-step system that made dumping "even the last of a three ton load" possible from street level. "This ease of loading, of course, greatly increases the number of collections daily per man." *Engineering News-Record* revealed its (and probably Vincenz's) racism in crediting further efficiencies to what it called "high-grade, white American labor" employed at the Fresno fill.[34]

Vincenz's retooled system was so cost-effective that his department reduced garbage collection rates for Fresno householders three times in the sanitary landfill's first three years of operation.[35] These inexpensive trash cemeteries and super-efficient collection methods represented the birth of refuse handling as an industry, functioning with assembly line ease, minimal labor and pared-down prices.[36]

Cover the Earth

The sanitary landfill was fully implemented in Fresno while other cities were struggling to develop acceptable burial methods. The perils of entombing trash persisted—there was as yet no uniform system of sanitary landfilling and many sites were hellish nests of fire and vermin. Although useful for waste disposal, packing the earth with trash was also increasingly viewed as a good way to reclaim what the *Engineering News-Record* called "hitherto unusable swampy and low-lying areas." This practice took off in the 1930s and 1940s as engineers created trash-based real estate from previously unoccupied tracts across the country.[37] As such, the sanitary landfill not only remade wasting, it also remade modern landscapes.

During the 1930s the benefits of filling in land with garbage were praised in places like New York's Jamaica Bay, a marshland that would otherwise "be useless for either industrial or recreational development."[38] Seattle built its thirty-five-acre Green Lake Playfield using trash to fill an "unsightly swamp."[39] And, with the help of rejectamenta, New Orleans hoped to see "old unused drainage canals and low places . . . transformed into beautiful parks, healthful playgrounds for children and good streets."[40] During these same years, Gordon M. Fair, a professor of sanitary engineering at Harvard, agreed that the sanitary landfill's possibilities were tremendous: "The city [rises] anew, like the Phoenix, from its own ashes: waste lands, unsightly and often offering breeding places for mosquitoes, [can be] converted into parks and attractive building sites. . . ."[41]

Despite official optimism, failures and disasters lurked at dumps operating under the new nomenclature of sanitary landfill. In the late 1930s, as reported in the urban planning journal *American City*, Oakland's bayside fill caught fire and "literally thousands of rats were smoked out and driven to other parts of the city." Not long after the blaze the War Department quarantined the port of Oakland for "unsanitary conditions on the waterfront."[42]

Dayton's landfill suffered overwhelming pest inundation. According to an article in *Civil Engineering*, nearby buildings swarmed with gray crickets, water bugs and sowbugs "so numerous as to obscure completely the sides of buildings. They are charged with eating the paper off the walls of houses and with devouring whole melons or whole cakes at night." Dousing the landfill with oil and setting it on fire to eradicate the devils proved unsuccessful, so dump managers blanketed the buried waste with "uncracked crude," which was "more or less effective but on account of the low flash point, many fires resulted, and dump fires are difficult to put out."[43]

San Francisco used landfilling extensively throughout the 1930s. After making pick-ups, local collectors dumped their haul at a railroad loading area at Sixth and Sixteenth Streets. There, unmixed restaurant slop was sold as hog feed and the remaining trash was packed into rail cars and sent to a bayside fill a few miles away. At San Francisco's marshy landfill, municipal castoffs were plunged deep into the bay mud at low tide.[44]

Using this method of landfilling, engineers conjured property out of San Francisco's previously uninhabitable southeast wetlands at a rate of one and a half acres every month.[45] As noted in *Civil Engineering*, in seven years, the city's sanitation department created about sixty acres of new land. San Francisco thus extended the peninsula 2,500 feet to the south and "easterly into the bay a uniform width of 1,050 ft. It is estimated that the refuse in the finished fill is approximately 50 ft. in depth, of which about 25 ft. is below the original mud surface."[46]

However, San Francisco's refuse depository experienced problems, also reported in *Civil Engineering*: "Wave action against the toe of the new fill caused sections of it to break away or settle, exposing faces of fermenting garbage and causing cracks through which foul gases escaped."[47] Regardless, the city was praised in the trade press and by engineers like Harrison P. Eddy, Jr., who made this creepy observation: "At very low tides during the winter, a black liquid occasionally is observed seeping out of the lower portion of the San Francisco fill. This seepage has an exceedingly disagreeable odor, but fortunately at that location it creates no nuisance, because it flows almost immediately into the bay waters."[48]

The sanitary landfill method was employed extensively in New York in the 1930s as well. Trenches dug into marshy land were packed with mixed discards and covered over with dirt or sand, then sprayed with a creosote solution commonly used to kill maggots. New York's first modern sanitary landfill was located on Rikers Island. So much waste was sent to the place that in 1938, after years of dumping, the island had almost doubled in size and towered 140 feet.[49] By the end of the 1930s, New York's landfill methods came under damning criticism, labeled by some as a "return to primitive methods."[50] The fill on Rikers Island was hardly the organized and sanitary disposal site envisioned by engineers like Vincenz. According to Thomas DeLisa, a sanitation department employee who worked on the island for eighteen years, the place was a horror:

> The rats became so numerous and so large that the department imported dogs in an effort to eliminate the rats . . . there were more than one hundred dogs on the island, dogs which were never fed by authorities, but lived solely on these rats. Despite this the rats, some of them as big as cats, continued to multiply. It was nothing to see as many as one hundred rats in a walk across the landfill at night. . . .
>
> Gases . . . were constantly exploding, erupting through the soil covering and bursting into flames. There was never a day in the summertime when fires were not breaking out and the stench from these fires gave off the most noxious odor imaginable. . . . When a hot spell would come along in the summer the ground resembled a sea of small volcanoes, all breathing smoke and flames.[51]

Meanwhile, New York's megalomaniacal parks commissioner, Robert Moses, was making use of trash to reclaim land all over the city. By filling in swampy wetlands with city wastes redirected from

Rikers Island and other municipal dump sites, the commissioner created vast tracts of new real estate for parks, roads, fairgrounds, airports and other facilities, doing much of this dirty work under the banner of his "Sanitation, Reclamation, Recreation" plan. In 1934 alone, Moses sent trash landfill crews to twelve different sites across Brooklyn and another nineteen in Queens, and plugged a 274-acre parcel on Staten Island to build a park. Some neighborhood groups who opposed Moses's unsightly, fetid dumping staged protests to express their anger over being excluded from the planning process, but they were ultimately unable to stop the commissioner.[52]

Moses's refuse-based projects included the Belt Parkway, Idlewild Airport (now John F. Kennedy Airport), Orchard Beach, Olmstead's Riverside Park along the Hudson, and most of the northern shore of the marshy Jamaica Bay, including Marine Park. Moses also decommissioned the city's trash processing hub on Barren Island and filled the wetlands around the island with garbage, fusing it to the mainland to conjure up what became Floyd Bennett Field, New York City's first municipal airport. Corona Meadows, the site of the 1939 World's Fair, was another Moses creation built on land filled in with trash.[53]

According to an article from the time in the *Engineering News-Record*, "The theme of the Fair was given expression in the phrase, 'Building the World of Tomorrow'—a slogan which for the first time in the history of expositions, looked to the future instead of the past."[54] Apparently, the future would be built on trash. And in light of Walter Benjamin's observation, "World exhibitions are the sites of pilgrimages to the commodity fetish," this was a monument to capitalism, with the dead commodities of the past literally serving as the foundation for the commodities of the future.[55] The World's Fair and its grounds were an omen of massive wasting to come.

Wartime Perfection

During World War II, the sanitary landfill was improved and standardized as a direct result of wartime strains on raw material supplies, the expertise of the U.S. military's Corps of Engineers, and the influence of the inimitable technocrat from California's Central Valley, Jean Vincenz. According to Daniel Thoreau Sicular in his investigation of the sanitary landfill's origins, Vincenz left Fresno in 1941 to serve as assistant chief of the Repairs and Utilities Division, War Department, Corps of Engineers, charged with maintaining utilities at domestic bases. To conserve materials the army restricted the use of incinerators and ordered that sanitary landfills be constructed on bases across the nine army regions in the United States. Because of his work in pioneering the method, the Corps put Vincenz to the task of writing a manual on how to build and run sanitary landfills. With the resulting manual as a guide, by 1944 over a hundred domestic bases were successfully disposing of their wastes at sanitary landfills.[56]

Throughout the early 1940s, Vincenz and other engineers learned how to design sanitary landfills more effectively, accounting for variables like climate and population size and making optimal use of space by maximizing compaction. The labor required for operating such sites remained minimal. A sanitary landfill, even for a large camp, could function properly with just one worker.[57] The Corps' engineers experimented with new and efficient equipment like the bullclam, which "could not only compact the refuse but also pick up and carry earth cover." Among the other emerging tools of the trash trade were recently invented front-end skip loaders, scrapers and tractors. These machines all made the job of burying refuse and smoothing it over under a peaceful layer of dirt and grass seed increasingly efficient and easy.[58]

From these military origins the sanitary landfill infiltrated all regions of the country. The technology spread via organizations like the U.S. Public Health Service, which in 1943 recommended the sanitary landfill to municipalities as a wartime labor and resource conservation measure. More significant were the many engineers who, after serving on domestic bases during the war, returned to jobs as teachers or in local public works departments. Freshly trained in building and maintaining sanitary landfills, these professionals had ready outlets for their new know-how. According to Sicular—who interviewed Vincenz in 1980—after the war these engineers, "who had acquired theoretical and practical knowledge of [the] sanitary landfill, as well as an appreciation for its benefits, returned to their old jobs. Many of them brought the sanitary landfill back with them, as a sort of domestic war souvenir. As Vincenz recalled, 'then the idea just caught on. It amazed me!'"[59]

The method received postwar support from other organizations like the powerful U.S. Chamber of Commerce's Health Advisory Council, which issued a brochure extolling the sanitary landfill's virtues and urging municipalities to take up the practice.[60] Business, it seemed, liked the way the landfill worked.

After World War II, low-cost landfilling beat burning hands down. In the 1950s the bill for operating an incinerator rang up at $2 to $6 per ton, whereas the costs of the sanitary landfill hung much lower, between $0.40 and $1.50 per ton.[61] By 1945, nearly one hundred municipalities had established sanitary landfills and within fifteen years 1,400 American cities were using the method.[62] According to *War on Waste*, written by environmental consultant Louis Blumberg and urban planning professor Robert Gottlieb, in the immediate post-war years, sanitary landfills "were already perceived as preferable to other forms of disposal for their ability to minimize the health concerns associated with open dumping and incineration, while utilizing engineering expertise in an era when technology was presumed capable of solving any number of social problems."[63]

Scavenging Transformed

The ever-expanding manufacturing sector discouraged repair and reuse of materials in the name of economic growth, even during the Depression. As the quantity of waste that Americans tossed out continued its industrially fueled upward spiral, and as collection and disposal grew increasingly consistent, scavenging became controversial and more tightly monitored. At the new sanitary landfill, gleaners were problematic because they got in the way and made a mess, driving up costs. Even when they were allowed to sift through refuse at the landfill, scavengers were often unable to retrieve sufficient usable materials; since the process required compression, items were typically crushed beyond repair. Nevertheless, salvaging and reuse did not fully disappear.

Industry and its spokespeople long feared that repair and reuse might lead to lagging consumption of new commodities, which could in turn foment a dreaded crisis of overproduction. In the 1930s, Richardson Wright, editor of *House & Garden*, advised:

> Saving and thrift would be the worst sort of citizenship today. . . . To maintain prosperity we must keep the machines working, for when machines are functioning men can labor and earn wages. The good citizen does not repair the old; he buys anew. The shoes that crack are to be thrown away. Don't patch them. When the car gets crotchety, haul it to the town's dump. Give to the ashman's oblivion the leaky pot, the broken umbrella, the clock that doesn't tick. . . . To maintain prosperity we must keep those machines going. Always we must be prepared to consume their enormous production.[64]

Informed directly and indirectly by this ethos, improved waste-handling methods and technologies actually engineered scavenging out. For operators like Vincenz, gleaning was incompatible with the cutting-edge dump: "It is my belief that if you wish to have a really sanitary fill it is difficult to carry on salvaging operations."[65] Initially, the Fresno commissioner had tried working with a junk dealer who paid the city a monthly fee to skim its discards. However, according to Vincenz the scavenger "soon had the site so cluttered with boxes of bottles, piles of cardboard and papers, and fenders, that I urged that his contract be canceled."[66] Vincenz then forbade salvaging at the Fresno landfill altogether.

In addition to logistical problems, aesthetics were a major issue for Vincenz. After all, the new sanitary landfills were highly self-conscious spaces—even the name sounded clean. Like Colonel Waring before him, Vincenz understood the power of aesthetic order and used public relations to promote his new waste disposal solution. He banned all fires from the site, planted trees, kept the place tidy, and invited journalists, schoolchildren and local officials to frolic on the sod-covered spoils.[67] Keeping garbage contained did cut down on pests, the noxious stench and stray trash that could blow about. In addition, camouflaging the dump with greenery cast wasting as benign while it categorized scavenging as unclean and therefore verboten.[68]

Los Angeles tightened its grip on its refuse operations in the early 1930s by declaring all discards the property of the city and adopting tough new ordinances outlawing the transporting of waste on city streets without a permit.[69] This made independent gleaning practically impossible. In 1934, as reported in the trade magazine *American City*, overseers of one of the Los Angeles incinerators discontinued "the old custom of allowing about fifteen junk salvagers . . . to operate on the charging floor picking paper, rags, cans, etc." As a result, "the hazard and unsightly piles of junk were eliminated."[70] Taking a less rigid approach, San Francisco let its contracted haulers cull what they wished while en route—on their own time—but no gleaning was allowed at the landfill.[71]

Even though public health departments frowned on scavenging, bans like those in Fresno, Los Angeles and San Francisco were not the norm. Some municipalities allowed salvaging out of concern for the spatial limitations of landfilling, while others tolerated it because so many people were in need during the Great Depression. Either way, forms of freelance and institutionalized waste picking persisted across the United States in the years leading up to World War II.

One official explained in 1938 that "if the typical American city were to adopt the sanitary-fill method of disposal today, it would not be many years before all the convenient areas would be filled up."[72] To cope, some municipalities started their own salvaging operations to conserve land disposal space. During the 1930s, Brooklyn opened a sorting plant at a new incinerator to rescue materials and thereby cut down on ash that needed to be hauled to the landfill. Using New Deal WPA funds, Milwaukee constructed a successful gleaning facility that extracted "enough salable material to pay much of the expense of the salvage plant . . . [and] enough combustible material . . . to heat several city buildings," all while cutting the city's ash output in half.[73]

As is true today, discarding was directly proportionate to the health of the economy; the fewer resources people had, the fewer substances they regarded as garbage.[74] During the Depression, municipal tolerance of gleaning was encouraged by the American Public Works Association: "Since the refuse contains articles and materials of value, needy persons should not be deprived of an opportunity to gain what they can through the recovery of materials that otherwise would be wasted."[75] According to the *Engineering News-Record*, in Seattle, "the contractor makes no attempt at salvage but allows pickers to work over the refuse and take out such items as they can sell."[76]

Meanwhile, groups like Goodwill Industries, the Salvation Army, and Saint Vincent de Paul contributed to a new kind of institutionalized salvaging. Founded in the previous century as religious charities, by the 1930s these organizations had a significant share of the salvage market with large operations collecting rags, paper and metal. They also solicited donations of unused and unwanted items like clothing and furniture for repair and resale. But while these groups supplied affordable commodities to those in need, crucially, they also kept the poorest engaged in the social relations of consumption. Shoppers had either to trade labor for goods or to purchase items with their own money; nothing was given for free.[77]

In addition, organizations like Goodwill facilitated the discarding of still usable items by the middle and working classes. Providing this outlet for castoffs helped construct wasting as acceptable—even pious—in many cases derailing tough questions about the ramifications of ever-expanding wasting.[78] If such items went to someone in need, then throwing things away must not be so bad.

World War II brought what would be the last great surge of reuse for decades. During the war, the country was faced with major shortages of raw materials. To address this problem, the government conducted scrap drives using eye-catching propaganda posters that urged citizens to turn in materials like aluminum, rubber and cooking fat for use in wartime production. While widespread scavenging ensued, a normalization of purging took hold as well. According to historian Susan Strasser, "Paradoxically, the very emphasis on scrap reinforced not the traditional stewardship of objects but the newer habits of throwing things away."[79]

These types of permitted and institutionalized salvaging took place in the context of an overall decline in the practice during the first half of the century. Reclaiming discarded goods no longer fit with the new sanitary disposal methods. In the 1950s mixed collection of discards became the norm across the country as mainstream Americans stopped separating their trash and started throwing more away. In lieu of salvaging, the sanitary landfill was perfect for managing the flood of trash that gushed from feverish postwar consumption.

Woman preparing to burn garbage in a backyard incinerator, ca 1950. (*San Francisco News-Call Bulletin,* courtesy of the San Francisco History Center, San Francisco Public Library)

5

The Golden Age of Waste

> Tomorrow, more than ever, our life will be "disposable."
>
> *Sales Management Magazine,* 1950

Perhaps no place or time was as ostensibly "clean" as 1950s America. From the clipped hedges and the sleek tailfins to dad's crew cut and mom's conical space-age bosom, the world seemed to be in perfect order. Central to this society-as-machine aesthetic was a plethora of new commodities made for easy use and quick disposal. This was the moment when the accumulated scientific breakthroughs of two massive world wars finally hit civilian life in full force. It was the age of the paper plate, polyester, fast food, disposable diapers, TV dinners, new refrigerators, washing machines and rapidly changing automobile styles. Most of all it was the epoch of packaging—lots of bright, clean, sterile packaging in the form of boxes, bags, cellophane wrappers and throwaway beer cans. The golden era of consumption had arrived, bringing the full materialization of modern garbage as we know it: soft, toxic, ubiquitous.

This new consumption-heavy reality was particularly fueled by the post–World War II boom, which radically transformed the production process and in turn led to both consumer bounty and a massive increase in waste. The advent of what is often called Fordism—mass assembly-line production coupled with mass standardized consumption and the mass psychology that goes with both—created a tidal wave of trash, and it did so in three distinct ways. First, the new Fordist economy used less recycled inputs from household wastes handled by independent junk traders; instead, scrap came from consortiums of large materials handlers. Second, it built new forms of waste into both commodities and the production process, in part because these cut labor time and externalized costs, thereby boosting profits. And third, the new economic regime created and demanded unprecedented levels of consumption, the main by-product of which was garbage.

Under this super-efficient postwar manufacturing system, America produced fully half of the world's wealth. But as the 1950s wore on, markets became increasingly saturated. There was more to the picture than just efficient manufacturing; commodities not only had to be produced—by now, that part was easy—they also had to be consumed, that is to say, destroyed. Within the first decade after World War II most consumers already owned what they needed, so how could industry sell them more? The answer was "built-in obsolescence." Producers began making commodities that intentionally wore out faster than was technologically necessary. And because of unprecedented production efficiencies, these commodities were becoming cheaper to replace than they were to repair.

Not surprisingly, refuse output was skyrocketing. By 1960, each American tossed out about two and a half pounds of trash daily.[1] Even though this was roughly equivalent in weight to waste generation in 1920, the contents in the national rubbish bin had dramatically changed. Minus the piles of ash, which constituted up to 75 percent of all wastes in earlier decades, trash in the postwar era was of an entirely new, abundant breed.

Because Waste Is Cheaper

During World War II, industry and workers were forced through government intervention to accept efficiencies at the factory level that kept output high. In doing so, manufacturers achieved levels of production that they had not reached before. And through mounting consolidation, concentration of the means of production in fewer hands helped create greater economies of scale—large-scale, more intense and efficient work regimes that ran around the clock, which increased productivity while raising profits and lowering commodity prices.[2] These transformations demanded greater quantities of inputs, sidelining smaller supply lines of reusables from sources such as household waste. It was also through employing such superefficient manufacturing methods that many of the costs of production—and consumption—were externalized onto the environment in new ways.

There were additional sources of waste associated with the new system of mass production. For one, Fordism relied on massive inventories in part because these gave employers greater leverage over striking workers. But these stocks were hard to match perfectly with demand and market conditions; if anything were to go wrong—a plunge in sales of *x*, a miscalculation about the supplies of *y*—massive amounts of goods and inputs instantly became trash.

Another emergent category of castoffs was rapidly outmoded factory equipment. Fierce competition compelled producers to constantly replace older machines that were often still functional.[3] For example, Andrew Carnegie obsessively upgraded the equipment in his U.S. Steel plants to achieve ever greater efficiency, since, as he put it, "We are bound to be followed very soon after we get started."[4]

In response to all this, a new kind of waste marketing arose from within the confines of industrial production itself. Corporations like Western Electric, DuPont and General Electric linked up in membership-based trade organizations that helped manage obsolete equipment or tons of wool, rubber or metal scrap collected off assembly line floors. One prominent such consortium, the National Association of Waste Materials Dealers (NAWMD), begun in 1913 and known today as the Institute of Scrap Recycling Industries, thrived with increased consolidation of industrial ownership.[5] Some of the wastes handled by groups like NAWMD were directed back into production and some were handed off for disposal. But the key impact on household refuse was that its pathways to the factory were diminishing. Consequently, consumer wastes were reused far less than in the past. In the years after World War II, household trash would most often be burned or buried.

The rationalization of industry that became Fordism was not simply a natural outcome of capitalist development. State policy helped shape this process in significant ways, contributing to the formation of today's high-waste system. In response to increased demands on resources during World War II, the federal government intervened in manufacturing through such agencies as the War Production Board (WPB), started just after Pearl Harbor to manage raw substances and oversee wartime commodity output.[6] The WPB restricted producers who used materials considered strategic, such as steel, paper, rubber and nylon, and shifted some factories from civilian manufacturing to building military equipment.[7] In implementing its controls, agencies like the WPB forced workers and owners to cooperate among themselves and with each other, creating unprecedented efficiencies on the factory floor.

According to urban geographer David Harvey, "War-time mobilization also implied large-scale planning as well as thorough rationalizations of the labour process in spite of worker resistance to assembly-line production and capitalist fears of centralized control. It was hard for either capitalists or workers to refuse rationalizations which improved efficiency at a time of all-out war effort."[8] The political alchemy created among the state, labor and capital during World War II thus produced a postwar environment that "brought Fordism to maturity as a fully-fledged and distinctive regime of accumulation."[9]

Streamlined industrial production entailed new ways of relating to waste. The efficiencies sought under postwar Fordist production, heavily influenced by Taylorist scientific management of the labor process, had everything to do with cutting out excess.[10] But the kind of "waste" these theories aimed to excise had little to do with trash. In capital's eyes waste referred to lost labor time during production—after all, surplus value generated by labor is the source of profits. In contrast, wastes that adversely affected the environment were largely treated as acceptable. From the castoffs of intensified raw materials extraction (such as mining slag) and factory wastes (like chemicals dumped into rivers), to the manufacturing of disposable goods deposited back into the environment after being discarded by consumers, mass production created an unprecedented surge in trash. In part, Fordist production drove down the price of manufacturing—and therefore of commodities—by externalizing these costs onto nature.

The new, improved, high-metabolic production lines were so successful that in the wake of World War II, as industry shifted from guns to butter to household appliances for both domestic and foreign markets, the rate of growth in per capita output more than tripled, and that for exports increased eightfold.[11] And of course, the amount of garbage grew proportionately.

The New Way

The post–World War II remaking of industry brought an analogous transformation of consumption habits and culture as America's middle class grew. At its heart, Fordism was marked by relative peace and cooperation among capital, labor and the state. The government regulated the economy, capital agreed to pay higher wages based on the greater productivity available with new manufacturing methods, and workers for their part accepted consumerism in place of political power on the job and in society as a whole. This new culture of wasting was founded on the increased income and welfare of large parts, though not all, of the working class. As wages steadily increased and commodity prices fell, consumption skyrocketed, and so too did its offspring, garbage.

These changes did not come seamlessly. The same Americans who released their legendary pent-up desire to consume in the years just after World War II were still accustomed to buying in bulk, stitching dresses from gunnysacks, and opening tin cans without electrical assistance—despite decades of industrialization. They were also fresh out of the Depression and a war economy within which conservation was highly valued. Slogans issued from the federal government's Office of Price Administration such as *"If you don't need it,* DON'T BUY IT!" and others like "Use It Up, Wear It Out, Make It Do, Or Do Without!" were still ringing in their ears.[12] But those years of material drought and deferred gratification converged with bulging savings accounts, a plethora of new commodities, and falling prices, unleashing a frenzy of postwar consumption.[13]

Between 1945 and 1950 consumer spending surged by 60 percent overall.[14] And in the four years following the war, Americans purchased 21.4 million cars, 20 million refrigerators, 5.5 million stoves and 11.6 million televisions, and they moved into over 1 million new housing units each year.[15]

The feverish postwar industrial output of commodities was a cultivated, nurtured outcome, rather than a natural by-product of capitalism. Processes encouraging mass consumption and disposal of such goods were devised and engineered with equal care, and strategic government programs were key components. To be sure, the public readily and enthusiastically took part in the postwar consumer mania, but not without the state's help in cultivating mass consumption and the culture of waste.

In the postwar years, government subsidies abounded in the form of FHA loans, the GI Bill and the Interstate Highway Act, among other.[16] All these financial inducements fostered the growth of a new American way of life. Suddenly, the purchase of a new car, a suburban home and trips to the shopping center were not only possible, but also sensible. Coupled with falling prices and low unemployment, these developments fueled unprecedented levels of individual consumption.

In 1959, Vice President Richard M. Nixon boasted, in his infamous "Kitchen Debate" with Soviet Premier Nikita Khrushchev, that each of the 44 million families in the United States had at least one car and one television set and the majority owned their own homes. In waging the ideological battle against the Soviets, Nixon argued that it was capitalism, not communism, that delivered the most goods to the most people: "Can only the rich in the U.S. afford such things? If this were the case, we would have to include in our definition of rich the millions of American wage earners."[17] As the Cold War gained momentum during the forties and fifties, government and U.S. industry alike began equating democracy with the freedom to purchase, creating a climate in which materialism and prodigality were recast as patriotic and wholesome.

Bursting at the Seams

All this stuff, both domestically and in foreign markets, began to pile up in living rooms, kitchens, garages and attics. To the alarm of manufacturers, by the end of the 1950s many U.S. families had already acquired an automobile and suburban home with all the gear they could possibly need.[18] What's more, the recovering production bases in Japan and Germany were beginning to compete with U.S. firms for what was turning into a shrinking market. The Fordist factory could crank out commodities like none other; manufacturing new goods to sell was not the problem. As the postwar period pressed on, the real trick was to keep people buying.

Paul Mazur, a Wall Street investor, outlined the problem in his book *The Standards We Raise*: "It is absolutely necessary that the products that roll from the assembly lines of mass production be consumed at an equally rapid rate. . . ."[19] The glutted markets of the late 1950s were giving rise to a potentially serious economic downturn. If left to its own devices, consumption would not keep pace with increased production. Industry had to act fast to head off this disaster.

One remedy was the product "line" or "family," a concept that dated back to the early twentieth century. This means of expanding consumption was applied with particular zeal by postwar manufacturers, creating a spectrum of commodities for every taste and budget. Existing products ranging from soap to refrigerators, radios and clothing now came in multiple colors, sizes and models, and consumers were encouraged to buy more than one of everything.[20] The head of refrigerator manufacturer Servel, Inc., informed consumers that the time had come for "two refrigerators in every home." A call also went out from an officer in the American Home Laundry Manufacturers Association for two washers and two dryers in each abode. And according to the Douglas Fir Plywood Association, "Every family needs two homes."[21]

But these appeals for voluntary over-consumption did not alleviate the looming crisis. In response, the supple minds of capital devised one of the most notorious market fixes to date—built-in obsolescence. Designing products with prefigured "death dates" allowed manufacturers to make commodities that intentionally wore out. These products became obsolete due to fashion and design changes, technological innovation, or lower production quality. Once outmoded or broken products were discarded, they could be replaced with new ones, solving the problem of slumping demand. As Vance Packard pointed out in his bestselling 1960 book *The Waste Makers*: "The way to end glut was to produce gluttons."[22]

In the nineteenth century, Karl Marx had already warned that new super-efficient industrialized manufacturing could easily outpace consumption: "The production of . . . surplus value based on the increase and development of the productive forces, requires the production of new consumption."[23] Back in postwar America, the chief of the J. Walter Thompson advertising agency understood this same threat: "We must cut down the time lag in expanding consumption to absorb this production."[24]

Built-in obsolescence was not such a new idea: its roots lay in the first half of the twentieth century.[25] Home economics writer and marketing consultant Christine Frederick was among the first to point out the virtues of what she called "progressive obsolescence." In the late 1920s, the free-market-loving Frederick explained that this type of wastefulness was not only inevitable but also life-affirming:

"It is the ambition of almost every American to practice progressive obsolescence as a ladder by which to climb to greater human satisfactions through the purchase of more of the fascinating and thrilling range of goods and services being offered today."[26]

After World War II, manufacturers applied the different types of obsolescence—technological or fashion—variously to stimulate market demand. Technological innovation could improve the function and efficiency of a commodity, but—crucially—it could alsobe wielded by producers to make goods that either wore out more quickly or were rapidly outpaced by improved models. Technological obsolescence was employed as early as 1939 when General Electric manufactured lightbulbs to burn out faster so they would need to be replaced more often.[27] But it was in the postwar period that technological obsolescence really took off.

In the 1950s, a Whirlpool engineering executive remarked that the company adjusted a product's "design-life goal . . . from time to time as economic or other conditions change."[28] Similarly, a Fairchild representative explained, "It is wasteful to make any component more durable than the weakest link, and ideally a product should fall apart all at once."[29] According to an ignition components manufacturer interviewed in 1960 by the *Wall Street Journal*, automobile manufacturers were building cars "so they'll get to the junk pile faster . . . today almost as soon as new cars hit the street they need replacement parts for all the gadgets they are loaded with."[30]

Fashion obsolescence was also extremely effective at compelling increased consumption. The automobile industry is illustrative terrain. The advertising journal *Printers' Ink* reported in 1959 that there was much more productive capacity than the market could absorb: "Our automobile plants could turn out eight million cars this year if they dared." However, the magazine explained, Detroit was only likely to sell half that number.[31] The auto industry needed to create markets by somehow convincing drivers to get rid of their still-functioning vehicles. They accomplished this by tapping into the psychology of aesthetic desire.

In the late 1950s General Motors famously set the industry abuzz with its plan to overhaul the entire body for each model every year, which had previously been prohibitively expensive. The coercive powers of competition ensured that other automakers followed suit. Around the same time, Ford hired George W. Walker, a "one time stylist for women's clothing," as its chief auto body designer. A Ford executive explained their new tactic: "The annual change cycle . . . is essential for competitive reasons. The change in the appearance of models each year increases car sales."[32]

Manufacturers of all kinds deployed built-in obsolescence with greater zeal throughout the 1950s, not only to get consumers to release their hard-earned cash, but also to get them to discard old possessions with ever-greater speed. Victor Lebow, an oft-quoted mid-century marketing consultant, explained the underlying logic:

> Our enormously productive economy demands that we make consumption our way of life, that we convert the buying and use of goods into rituals, that we seek our spiritual satisfactions, our ego satisfactions, in consumption. . . .
>
> We need things consumed, burned up, worn out, replaced, and discarded at an ever increasing pace.[33]

Urban geographer Richard Walker describes built-in obsolescence as "another aspect of capitalism's creative destruction. It isn't just tearing down factories or tearing down places, but it's actually tearing down its own products."[34] Ethereal needs linked to desire in tandem with other forms of technologically based superannuation drove greater levels of consumption, successfully rekindling the postwar economic fire.

Some engineers, manufacturers and journals grappled with the ethics and potential consumer backlash of making products designed to break or become outmoded at heightened rates. The trade publication *Design News* presented the question only to dismiss it with an irreverent answer that reflected the thinking dominant in U.S. industry: "Is this concept bad? We don't think so."[35]

It's So Easy

The apex of built-in obsolescence was the disposable commodity. Marketed under the alluring dual banner of cleanliness and convenience, widespread production and consumption of disposables kicked off a whole new level of wasting. Although disposables had been around for decades—the nineteenth-century throwaway, cork-lined tin bottle cap, designed by William Painter, was among the first—it was not until the postwar era that single-use products captured a sizable share of the buying public's dollars. Chief among disposables was the swelling category of product packaging. Although some of these ready-made nondurables might be reused before getting chucked, in the end more stuff was destined for the garbage graveyard.

New discardable products were just as mesmerizing as they were numerous. And while many of them undoubtedly made people's lives easier, single-use items comprised a plentiful new category of trash. Proctor & Gamble introduced the first disposable diaper, its Pampers brand, in 1961. Within a few years, Playtex started selling its Dryper disposable diaper, which, according to a TV ad, made "traveling with baby more fun."[36] After introducing the disposable razor blade in the early twentieth century as a replacement for reusable straight razors, Gillette would go on to refine its throwaway blade design, helping to acculturate a new generation to everyday disposability.[37] Also at the time, the Aluminum Company of America marketed new single-use cooking containers—no fuss with washing, just throw out the dirty pan.[38]

Mass desuetude hit new heights with disposable packaging. As mom-and-pop stores with their advice-giving sales clerks gave way to self-service chain supermarkets, the package became the producer's "sole representative at the sales decision point."[39] Packaging, which had previously been a subset of manufacturing, now became a subset of advertising, crucial to seizing the customer's attention and compelling buyer loyalty. And shoppers responded: good-looking packaging clearly enhanced the pleasure of consuming. Disposable wrappers and containers proved so successful a marketing strategy that by the 1960s industries were for the first time spending almost as much on packaging as they spent on manufacturing and traditional advertising.[40]

In the postwar years, throwaway containers crowded out bulk bins and "dual-use" packaging like tobacco tins that could double as lunch boxes.[41] In a few short years, Standard Packaging expanded from making disposable milk-bottle tops to manufacturing a huge range of discardable trays, boxes, bags, plates, bowls, utensils and "flexible packaging material." The company almost tripled its sales between 1955 and 1958. Its head, R. Carl "Hap" Chandler, explained why: "Everything we make is thrown away."[42]

Disposable packaging meant multiple successes for producers. Not only were companies more actively pushing their products, they also created a significant new revenue stream where these commodities—barely perceptible as commodities—were quickly and easily trashed, making room for still more consumption. Individual shoppers paid for the increased expense of packaging, contained as it was within the price of the product, and were left to fund the management of wastes themselves. In that scenario, still true today, the expense of packaging was externalized off the ledgers of industry and onto the bankbooks of consumers and taxpayers.

The annual cost for packaging in the 1950s was $25 billion. That meant that each U.S. family was paying $500 a year for packaging alone—a price that did not include municipal disposal or long-term environmental costs.[43]

With trash handling now under the attentive domain of municipalities, obsolescence and disposability appeared manageable; refuse was easily left at the curbside, spirited away and never seen again. Characteristic of modernity, the shift to carefree wasting was "the means to greater freedom from the shackles of tradition, from outworn equipment and ideas."[44] Liberated from repair, reuse and the tending of natural systems, the consumer was now free to waste.

Plastics

The wastefulness of the postwar period was embodied most fully in plastics. Synthetics entered the daily lives of Americans in the form of cheap commodities and packaging, both geared toward disposability. In those critical years after World War II, plastics production grew at a rate of more than 15 percent, almost quadruple that of steel.[45] By 1960 plastics surpassed aluminum to become one of the largest industries in the country.[46] Polymers were so popular in large part because manufacturers loved them. As the industry journal *Modern Plastics* pointed out, this inexpensive raw material was designed "by man to his own specifications" and so it could facilitate an unfettered production flow.[47] Beginning in the mid-1930s, switching to plastics proved extremely efficient: the same worker who had turned out 350 hair combs per day could now make more than 10,000 in equal time.[48] In the coming decades, with the help of government contracts and funding (especially during World War II), resin making blossomed into a major U.S. industry, and an abundant, toxic wellspring of garbage.

Invented in the nineteenth century as a replacement for raw substances like ivory, rubber and shellac, plastic was originally conceived as the remedy for restricted and dwindling natural resources. From the beginning, synthetics promised to cut manufacturers free from one of the greatest obstacles in industrial production—the limits of nature. In this regard, polymers also had a spatial component. No longer would producers have to venture across the globe for particular raw materials; now human-made substitutes could be cooked up in onsite laboratories as needed. Indeed, the key to plastic's success was that it possessed the flexibility demanded by a free market. Under capitalism anything can be commodified—common grazing fields one day, emotional services the next—and plastic, with its immediacy and infinite elasticity of forms, suited this economic system perfectly. As Roland Barthes noted, with plastic "the hierarchy of substances is abolished: a single one replaces them all: the whole world *can* be plasticized."[49]

With the deep pockets and industrial oversight of the U.S. government, the plastics industry matured during World War II and flourished in its wake. A few months after Pearl Harbor, a powerful industry group, the Society of Plastics Industries (SPI), sent board member William T. Cruse to Washington to work as a full-time liaison with the Office of Production Management (OPM), the body in charge of safeguarding industry against raw materials shortages during the war.[50] As editor of *Modern Plastics* magazine and former sales director for the Celluloid Corporation, Cruse was a plastics booster par excellence. Just a year after the industry's man hit the capital, the director of priorities for the OPM stated, "The impact of the rapidly developing defense program on our economic system now makes it imperative that certain vitally essential metals be conserved for primary defense purposes. . . . This means that the whole question of plastics now becomes more important than ever before."[51]

On factory floors and in the chemistry labs of the biggest plastics producers a revolution was soon under way. Used to make molded gunner's enclosures, cockpit windows, mortar fuses, helmet liners, goggles, raincoats, waterproof tents, parachutes, color-coded electrical wiring, and parts for the atomic bomb, plastics seeped into all levels of military equipment.[52] Dow even deployed its new Saran film to protect entire airplanes, artillery and other "sensitive military equipment" from salt and sea spray during transatlantic shipping.[53]

To meet a congressionally mandated production goal of 800,000 tons of synthetic rubber by 1944, the federal government threw down $1 billion for private companies to construct plants in cities from Louisiana to Connecticut. The target output was quickly met, proving that manufacturing fake rubber could massively outpace previous levels of refining "crude" natural rubber.[54] Synthetics were becoming a mass-production success.

As in other fields, government oversight drove unprecedented industry-wide cooperation. Consequent information sharing and standardization led manufacturers to develop more refined materials and perfected production processes.[55] During this time, chemists engineered plastic to exact molecular specifications that, according to *Harper's*, allowed a producer to "make a list of the properties he would like embodied in a new material" and then "custom-build the material as he never could before in all history."[56] On the factory floor, the fine-tuning of a process called injection molding—shooting liquid plastic into premade molds—was another major turning point. In concert with new synthetics mixtures, the improved injection molding production line made real every manufacturer's dream: the possibility for "continuous mass production."[57] As the company publication *Monsanto Magazine* gleefully proclaimed about injection molding manufacturing, "The single unit production line spews out the articles faster than we can tell you about it."[58]

Thanks in large part to significant government investment, annual U.S. plastics production *tripled* between 1940 and 1945.[59]

Immediately after the war, according to one executive, "Virtually nothing [for everyday consumers] was made of plastic and *anything* could be."[60] Emerging from the hothouse of wartime government contracts capable of unprecedented output levels, synthetics factories were erupting with Tupperware, Formica tables, Fiberglas chairs, Naugahyde love seats, hula hoops, disposable Bic pens, polyester leisure suits, silly putty, and nylon pantyhose. Not long after the war a manufacturer in Fort Worth, Texas, made the first plastic garbage can; at twenty-two gallons it was the largest injection-molded item yet made.[61]

But as markets became saturated, the polymer industry had to seek out new ways of expanding and intensifying consumption. A speaker at a 1956 SPI conference announced the remedy to his polymer-producing audience: "Your future is in the garbage wagon!" He urged the attendees to aim at "low cost, big volume, practicability, and *expendability*."[62] Built-in obsolescence was the order of the day; if consumers threw out more, producers could sell more. As a *Modern Plastics* editorial explained that same year, "In a prosperous consumption economy the factor of disposability is an important key to continuing volume."[63]

There were countless synthetics now manufactured to be disposable: diapers, plates, cups, lobster bibs, cabinet shelf lining, film for storm windows, and medical exam gloves.[64] Foster Grant, a plastics molding company from Leominster, Massachusetts (home to many resin makers), started up its own line of sunglasses: they were affordable and their styles were frequently changed, so consumers were enticed to toss out their old shades regularly for a new look.[65]

In addition to fashion obsolescence and single-use goods, packaging presented a ripe opportunity for boosting polymer consumption. A range of synthetics was employed for new disposable containers. Used during the war for flotation devices, Styrofoam emerged on the postwar disposables market most notably in the form of coffee cups and packing materials, and later as egg cartons, meat trays and take-out food containers.[66] Tastee-Freez, "with hundreds of roadside stands throughout the country," started dishing up ice cream in disposable styrene bowls.[67] Colorful, eye-catching and squeezable, polyethylene quickly became a popular packaging material.[68] "All those bottles on the shelf!" read one plastics ad from the mid-1950s, "Yet only one shouts 'reach for me!'"[69] By the 1960s polyethylene was used to make all order of disposables, from bleach and detergent bottles to lids on coffee and shortening cans, squeeze tubes for suntan lotion, bread bags, cosmetics packaging, and beer six-pack connectors.[70]

After the war, resin makers had mounted a major educational effort to accommodate the consumer to new, previously unknown plastics. People neither naturally gravitated to the stuff, nor did they instinctively throw it away, so the industry also had to inculcate consumers to plastic's disposability. In the late 1950s, according to a *Modern Plastics* article, even though people still tended to save and reuse new single-use plastic drinking cups, "It is only a matter of time until the public accepts the plastics cups as more convenient containers that are completely discardable."[71]

That shoppers had to be taught to consume and discard synthetics illustrates that the ever-expanding plastics market was not simply the result of consumer demand for convenience, as is often argued by the industry. As producers switched to synthetics, consumer choice for other substances was narrowed.[72] When manufacturers embraced plastic's versatility, low cost and distinctive look, the availability of products made with glass, paper and other natural materials began to dwindle. Plastics meant greater profits, and industry obeyed the economic incentive to switch. In this regard, American shoppers had no say in decisions made further up the production chain. In 1960 about 10 percent of all polymers produced went into packaging; by 1966 that number had doubled, and by 1969 nearly a quarter of plastics were made into packaging, almost all of which eventually became garbage.[73] So, while some supermarket-goers may have welcomed the new wonder substance, plastic and disposability were increasingly what was for sale, regardless of consumer demand.

Hidden Persuasion

Since all this consumption and wasting was not simply the normal outcome of democracy, markets and human nature, such behavior had to be cultivated. As the advertising journal *Printers' Ink* had noted before World War II, "The future of business lay in its ability to *manufacture consumers* as well as products."[74] To do so industry promoted individual, domestic consumption through advertising, often focusing on what some referred to as "invented" or "artificial" needs.[75] Spending on advertising mushroomed from $1 billion in

1920 to more than $4.5 billion in 1950, and by 1956 expenditures doubled again to almost $10 billion.[76]

Much of this appeared on the new and bewitching advertising medium of television. Built into the living room walls of Levittown's suburban dwellings as early as 1950, the television was becoming a standard household appliance.[77] More than 35 million families were glued to the tube by the mid-1950s.[78] Correspondingly, between 1950 and 1955 spending on TV advertising jumped from $177 million to over $1 billion.[79]

Postwar advertising grew out of marketing practices that targeted the individual. Incipient efforts at advertising in the late nineteenth and early twentieth centuries had highlighted the production side of the commodity to convince the buyer that the factory was clean, the ingredients unadulterated, and the company upstanding.[80] By the 1920s, producers were looking for novel ways to sell their wares, so many advertisers began generating new types of consumer needs rooted in human psychology. Constructing the consumer as plagued by inadequacies and threatened by his or her own failings, these advertisements tapped into insecurities by exploiting "problems" like bad breath, body odor or lackluster sartorial taste.[81]

Vigorously deployed in the postwar era, marketing based on desire, anxiety and envy was highly effective. This strategy produced a consuming class that did not look to the structural problems of industrial society as the source of its ills, but instead turned to industrially produced commodities as the solution.

Since well before World War II, as Elizabeth and Stuart Ewen have shown, advertising has reconstructed the worker not as part of a self-aware class but as an atomized "employee." This kind of advertising connected social status and human value with the ability to consume. In doing so, such marketing segregated consumption from the labor process, helping to construct an individual who was hailed in the marketplace not as a worker, but as a consumer. And within this structure the individual could express agency most powerfully at the point of purchase instead of in unified actions with fellow workers.[82] Here the laborer was seen as disempowered and impotent while the consumer was portrayed as effective and in control. Under these conditions, as John Berger has pointed out, the individual "lives, continually subject to an envy which, compounded with his sense of powerlessness, dissolves into recurrent day-dreams. . . . In his or her day-dreams the passive worker becomes the active consumer. The working self envies the consuming self."[83]

According to the Ewens, "Ultimately within a rising universal marketplace, *consumerism* is the basic social relationship replacing customary bonds."[84] Even bonds between humans and ecological systems were recast according to the contours of the market. The construction of the consumer as detached from both work and nature helped foster widespread acceptance of a refuse-heavy reality.

Out with the Old

Just as producers were creating expanded consumption using built-in obsolescence, waste handling had to evolve to accommodate the resulting maelstrom of refuse. In addition to the hyperefficient, widely used sanitary landfill, many other trash technologies proliferated after World War II. Exotic devices like one specially designed car incinerator called the "Smokatron," which could "process two car bodies an hour," were joined by innovations like new high-tech refuse compaction trucks.[85] Also, the postwar household now came equipped with up-to-date refuse treatment appliances. All these newfangled contraptions helped ease the practical, aesthetic and psychological burden of massively increased wasting.

With electricity and gas now coursing through most American homes, the amount of household ash waste decreased, but this change also put an end to the daily burning of paper and food scraps in the kitchen stove. To cope, some homeowners turned to in-sink garbage grinding: installed under kitchen drains, the units pulverized food wastes mixed with water, then washed the slop into local sewage systems.

Invented by Wisconsin architect John W. Hammes, the first kitchen garbage grinder, called the "In-Sink-Erator," was initially marketed in 1937.[86] Also referred to as the "disposall," these devices were banned by some local governments, like New York City's, for fear of overtaxing aging sewage systems. Similarly, some building owners avoided in-sink grinders to protect older plumbing.[87] Nevertheless, such "disposalls" were popular among many local waste authorities because they reduced the amount of garbage that needed to be collected, thereby lowering hauling fees. Endorsed by the American Public Works Association, over 2 million household garbage grinders were installed during the 1950s, with some cities actually requiring the gadgets in newly constructed abodes.[88]

Small-scale incinerators were also widely employed for at-home disposal. These furnaces were used in and outside of houses and apartment buildings, as well as stores, small industries, hospitals, and other institutions "to burn refuse produced on the premises . . . almost as soon as it is produced, reducing nuisances and hazards from it."[89] Such was the home marked by bomb shelter logic.

Another major postwar innovation, still the norm today, was the compaction garbage collection truck. These vehicles were designed specifically to handle greater volume while controlling labor costs.[90] A technology compatible with the sanitary landfill, the compaction truck hydraulically compressed all refuse together inside its enclosed wagon. By condensing its contents and further automating loading, the new truck required fewer workers and allowed a single vehicle to collect more refuse per trip, simplifying and speeding the costliest part of the trash handling process. Delivering precompacted trash to sanitary landfills worked like a charm.

As a by-product, all these new technologies diminished scavenging further. With in-house disposal methods, wastes never made it to the hands of gleaners. And since everything was crushed in the compaction truck's hull, salvaging, even by collectors, was made more difficult than before. So efficient at cutting costs and so harmonious with new streamlined refuse handling, by the mid-1970s compaction trucks comprised over half the collection vehicles in operation in the United States.[91] The cleanliness and immediacy of these waste-handling methods ultimately worked to help keep the commodities flowing; when wastes were easier to get rid of, through such an efficient infrastructure, the throwaway mentality seemed so right.

Woman being ticketed for garbage infraction, 1969. (Courtesy the New York City Municipal Archives)

6

Spaceship Earth: Waste and Environmentalism

An unprecedented popular uprising burst forth on April 22, 1970, when 20 million Americans poured into streets, parks and schoolyards across the country, proclaiming their concern about a new political problem: the environment. Organized by Senator Gaylord Nelson and activist groups like Environmental Action using the successful model of antiwar protests and teach-ins, the first Earth Day was a milestone. Ostensibly it was about the health of life on the planet, but at its roots Earth Day was the first popular challenge to America's wasteful production system. In towns across the country, people discussed the plight of the environment in an industrial world and took part in educational and political events reflecting regional concerns. Endorsed by Interior Secretary Walter J. Hickel, the event attracted children, mothers, farmers, ministers, historians, teachers, politicians and hippies alike. Even President Nixon, whose White House kept a cool distance, said that he approved of the event.[1]

In a speech that day, Chicago Seven member Rennie Davis declared, "Yes, it's official—the conspiracy against pollution. And we have a simple program—arrest Agnew and smash capitalism. We make only one exception to our pollution stand—everyone should light up a joint and get stoned." Those more moderate also offered interpretations of the ecological crisis, many condemning business as usual. Congregationalist minister Channing E. Phillips announced, "Environmental rape is a fact of our national life only because it is more profitable than responsible stewardship of earth's limited resources." Walter S. Howard, a biologist, told the crowd, "The affluent society has become an effluent society. The 6 percent of the world's population in the United States produces 70 percent or more of the world's solid wastes."[2]

Students at California's San Jose State University bought a brand-new $2,500 car and buried it as a protest against consumerism.[3] A shrimp trawler set out from Charleston, South Carolina, to Washington, D.C., to deliver a petition signed by 35,000 residents opposing a chemical plant planned for construction on the state's coastline. In the Motor City forty women marked Earth Day by picketing the Great Lakes Steel Corporation for dumping industrial wastes into the Detroit River. And young people collected garbage along U.S. Route 54 in Lake Ozark, Missouri, leaving "five piles more than ten feet high along the roadside." On the legislative front, Senator Nelson proposed a constitutional amendment that would give citizens new legal rights to demand environmental reforms.[4] And that day, masses of people began chanting a new mantra: "Reduce, Reuse, Recycle!"

Earth Day was not the moment that triggered environmentalism; rather it was the culmination of a series of events that directly and indirectly revealed modern production's dangers. The proof that all was not right had been mounting: Long understood as insalubrious and aesthetically displeasing, air pollution made headlines everywhere when, in 1952 London, toxic factory smoke trapped by a thermal inversion, which effectively formed a cap over the area, killed more than 4,000 people. Similar meteorological events led to two separate incidents of massacre-by-smog in New York City a decade later, leaving almost 600 dead.[5] On another front, Rachel Carson's 1962 book *Silent Spring* explained in lay terms the shocking effects of the widely used pesticide DDT, warning of a future plagued by toxic death and devastation. The bestseller ignited an uproar from the previously uninformed public and a PR backlash against the author by Monsanto and other chemical companies.[6] During the 1960s public involvement in fighting projects like dams and the logging of old growth forests pressured the federal government to pass the 1964 Wilderness Act, marking a new "ecology movement."[7]

It wasn't just environmental issues that pushed growing numbers toward heightened awareness of the problems with the status quo. The 1960s counterculture shunned conventional lifestyles and the superficiality of consumerism, with many of its adherents opting instead to live cheaply and scavenge the discards of an affluent society. In the latter years of the decade, riots rocked U.S. cities while political uprisings around the globe in 1968 ruptured the placid façade of the postwar years. And *Apollo* 8's voyage to the moon, just months before the inaugural E-Day, produced one of the first photographs of planet earth from outer space. This image unexpectedly showed viewers everywhere an image of the earth as a finite resource and helped trigger an awareness of the limits of nature's abundance, creating what came to be known as the "overview effect."

So prominent had the ecology issue become that Nixon felt obliged to address the subject in his 1970 State of the Union speech: "The 1970s absolutely must be the years when America pays its debt to the past by reclaiming the purity of its air, its waters and our living environment. It is literally now or never."[8] Such extreme talk from the U.S. president could only mean that the times were most definitely a-changing.

The burgeoning environmental movement revealed the widening gulf between a public increasingly aware of the limits of nature and an industrial sector hell-bent on ever-expanding production no matter what the ecological toll. As defined by most manufacturers, waste was whatever slowed down production. For the millions in the streets that first Earth Day, waste was barren forests, poisoned waterways, choked skies and endless piles of trash.

More people understood that the postwar era of prosperity had come at a price, that the apparent successes of streamlined production, massive growth rates, high profits and booming output levels actually resulted from a system that factored *out* the impact of its enterprise on natural systems. Taylorist, Fordist and the emerging postindustrial forms of production all externalized environmental costs, and environmentalists realized that this bill would inevitably come due.

Until the 1970s, the U.S. government had done little to counteract industry's displacement of costs onto ecosystems and human health. In fact, the state massively assisted companies in more fully exploiting natural resources. On the "back end," wastes—all those spent appliances, outmoded fashions, single-use containers—were treated as a public responsibility, with municipal agencies overseeing and financing technologies and programs to manage them. On the "front end," scant legal restrictions and an array of subsidies for raw materials extraction and transportation supported forms of production that devastated natural systems.

But the era's forceful political and cultural shifts pushed forward a wave of federal reforms that, while seen by some as timid and inadequate, marked the most aggressive attempt by the state to regulate the use of natural resources to date. In 1970, on Nixon's watch, Congress passed sweeping environmental protection laws including the Clean Air Act and the Resource Recovery Act. That year federal policymakers also created the Environmental Protection Agency with a mandate to "prevent or eliminate damage to the environment."[9] The Clean Water Act was enacted in 1972 and, in 1976, the Resource Conservation and Recovery Act was passed, requiring the EPA to oversee the development of landfill standards; this "marked the federal government's first involvement in regulating the disposal of hazardous waste."[10]

The Sierra Club and other old-line conservation groups who lobbied in Washington, D.C., were joined by newer, less staid organizations like Friends of the Earth, Environmental Action, the Environmental Policy Center, the Institute for Local Self Reliance and the Citizens' Committee on Natural Resources.[11] Also established in the early 1970s were Ralph Nader's Public Interest Research Group and the Natural Resources Defense Council. The newer organizations were formed out of public concern over pollution while the eco-establishment groups had come together in the early part of the twentieth century as conservationists. The legacy of the old-line environmentalists was the national forests; the agenda of the young activists was reforming the system to protect the health of humans and natural systems. In the 1970s many of these organizations went toe-to-toe with politicians and deft, deep-pocketed lobby groups including the National Association of Manufacturers (NAM) over issues like restrictions on packaging. These proposed regulations, known as "bottle bill" and "ban the can" laws, ranged from deposit requirements to outlawing certain types of packaging and led to a string of fierce legislative battles between environmentalists and big business.

No-Deposit, No-Return Blues

Beverage containers comprised the fastest growing component of solid waste by the mid-1970s.[12] Between 1959 and 1972, while the quantity of beer and soft drinks consumed increased 33 percent per capita, the number of containers consumed skyrocketed by 221 percent.[13] And with minimal restrictions on production and disposal, all those barely used bottles and cans were going straight to the garbage pile; by 1976 packaging, measured by weight, had become the single largest category of municipal solid waste, at 34 percent.[14] It made sense that environmentalists were suddenly obsessing about packaging. Before the 1970s most beverage manufacturers delivered their product in thick refillable glass bottles that could be washed and reused twenty times or more.[15] But the employment of refillables for soda plummeted from 98 percent in 1958 to just 39 percent in 1972. By that same year, 82 percent of containers for beer were "one-way."[16] A new garbage era had arrived.

The industry consistently claimed, as it does today, that the switch to disposables was the result of consumer demand for convenience—with throwaways those pesky, time-consuming trips to return empties became obsolete. In reality, this shift was the product of three main thrusts: beverage and container producers recognized the new, potentially enormous trade in one-ways; unique forms of packaging allowed for innovative marketing; and, perhaps most significantly, disposables were key to the consolidation of the beverage industry.

Although the first single-use can and bottle were introduced in the 1930s, disposables did not capture a sizable portion of the market until after World War II for various reasons. The two most prominent were that producers had yet to fully understand the profit potential of throwaways, and wartime restrictions on metal and glass kept mass production of disposables at bay. All that changed in the years after the war. With the hunt for ever-expanding markets heating up, executives in the drink and container industries, like their counterparts in most other fields, jumped on the garbage gravy train.

The profit potential with throwaways was staggering: after each disposable container was used only once it was discarded, so for every reusable bottle there could be anywhere from twenty to forty single-use containers consumed and permanently trashed. As the industry magazine *Modern Packaging* noted in 1961, "There appears to be a gigantic field for growth in non-returnables."[17]

The advantages of throwaways for container makers were obvious—more units trashed meant room for more new consumption. But disposables also presented marketing benefits. By selling beer and soda in novel one-way vessels, beverage companies (like all other manufacturers at the time) could grab consumers' attention, differentiating their brew from those of rivals on crammed self-service supermarket shelves. Known in the industry as "nonprice product promotion," beverage makers could now use flashy packaging instead of lower prices to get an edge on the competition.[18] New trashable wrappers allowed for expanding market share while keeping prices the same, or even raising them.

In addition to these developments, disposable packaging played a central role in the dramatic restructuring of the beverage industry in the years following World War II. By replacing refillable bottles with throwaway cans and glass containers, the major drink makers were able to more easily consolidate the beer and soda markets. Like other fields at the time, the beverage industry was becoming increasingly centralized and streamlined. In 1974, a local Pennsylvania bottler, Peter Chokola, explained that reusable containers imposed "a natural limitation on the market area served by a bottling plant." Delivery trucks were tethered by the refillable system; they could not venture long distances because they had to return to the bottling plant with the empties. According to Chokola, "Thousands of small and medium hometown bottling plants were therefore necessary to market beverages." Switching to disposables allowed large beverage makers like Coca-Cola and Pepsi to bypass smaller producers and, as Chokola described, throwaways "provided the medium through which monopolization of the soft drink industry could be achieved."[19]

As the returnable bottle system was being scrapped, thousands of local and regional bottlers began folding. While there were 5,200 soft drink makers in the United States in 1947, by 1970, just 1,600 remained. Consolidation in the beer industry was even more extreme. In 1950 the United States had more than 400 breweries, but within sixteen years that number had dwindled to just over 100. By 1974 there remained only 64. A study conducted for the EPA said that monopolization in the brewing industry was "encouraged and permitted by the introduction of nonreturnable containers."[20]

Along with these shifts in production, consumption and wasting practices transformed. As the expanding national brewers drained the regionals' business and as suburbia flourished, individually bottled six-packs designed for take-home consumption replaced old-fashioned keg distribution to taverns.[21] Additionally, one-way containers were favored by emerging grocery store chains because the disposables demanded neither the added labor time nor the valuable storage space that reusables required. These changes meant gargantuan increases in wasting. And by switching to disposables, industries connected to the beverage market furthered the Fordist practice of externalizing costs—onto the consumer, onto municipal refuse collection and disposal, and onto the environment. An *Antitrust Bulletin* article from 1975 called this externalization "private-to-social cost shifting."[22] These transformations—consolidation, wildly increased packaging consumption and the displacement of costs—all unfailingly pushed profits ever higher.

Welcome to the Biosphere

Packaging and disposables came under increased scrutiny as more people started questioning mainstream conventions. In the years leading up to the first Earth Day the counterculture was thriving as anticonsumerism flared almost unexpectedly from the heart of the American free market. Scaled-back consumption, volunteer recycling programs, and individual efforts at salvaging and reuse were seen by some as central in responsibly managing the planet, and avoiding the trap of endless work and debt.

Across the country long-haired dropouts culled supermarket dumpsters for free food and combed thrift stores for reusable clothes. Underground and subversive presses helped form nationwide resource networks and offered advice on alternative ways of living.[23] Abbie Hoffman's *Steal This Book* gave instructions on where to find free stuff:

> Landlords renovating buildings throw out stoves, tables, lamps, refrigerators and carpeting. In most cities, each area has a day designated for discarding bulk objects. Call the Sanitation Department and say you live in that part of town which would be putting out the most expensive shit and find out the pick-up day. Fantastic buys can be found cruising the streets late at night. Check out the backs of large department stores for floor models, window displays and slightly damaged furniture being discarded.[24]

Hippies and bohemians baked bread, threw clay pots, built their own furniture, sewed their own clothes, grew food in urban gardens and pooled resources in communal houses. The more stalwart went "back to the land" to live without electricity or running water. New cooperative food stores like Rainbow Grocery in San Francisco and Stone Soup in Washington, D.C., avoided selling overpackaged food by stocking their supplies in bulk bins and asking their customers to bring their own bags.[25] Thus the counterculturists and those they influenced reduced consumption and made and repaired the things they used in their daily lives, creating pockets of resistance to the terms set by U.S. industrial manufacturers.

Also during this time, ecologically minded volunteers started neighborhood recycling programs for paper, cans and bottles. Not since the resource austerity of World War II with its nationwide scrap drives had any major effort at recycling been made in the United States. Madison, Wisconsin, became the first municipality to implement postwar curbside recycling collection (for newspapers) in 1968 and just six years later, due to public pressure, more than 130 municipalities offered some kind of curbside collection program. In the months straddling the first Earth Day as many as 3,000 neighborhood recycling programs started up.[26]

Volunteers—many of them women—organized these early recycling efforts, first out of their kitchens and garages, then out of neighborhood drop-off centers.[27] According to historians Louis Blumberg and Robert Gottlieb, such programs "were initiated not just as a way to establish a solid waste management alternative, but as a kind of cultural rejection of the 'throwaway' market."[28] Many of these recycling depots became information centers, offering an array of environmental facts that addressed the impacts of production and consumption, which up until this point had largely remained hidden.

Waste-mitigation methods forged by the eco-vanguard began capturing the interest of more mainstream institutions. For instance, the EPA borrowed from these practices to formulate what it called "source reduction." In the mid-1970s Eileen Claussen of the EPA's Office of Solid Waste Management explained the concept: "Most approaches to waste management have dealt with the means of disposing wastes without considering the environmental pressures generated in the creation of wastes. In contrast, source reduction would decrease environmental pressure from the time of the extraction of raw material through the final disposal stage."[29]

Crucially, these more holistic approaches viewed the "three Rs" in a hierarchical fashion: first, reduce consumption, then reuse goods in their already manufactured form as long as possible, and then, only as a last resort, recycle. As explained in 1974 by Environmental Action's solid waste coordinator, Patricia Taylor: "Ideally, every method of waste reduction should be explored before resource recovery [recycling] methods are implemented. Taxpayers should not finance resource recovery facilities designed to handle materials that need not have been produced in the first place."[30]

Getting at the causes and not just the symptoms, hippies, activists and other concerned citizens began calling for greater reuse and increased product durability and serviceability, to reduce materials and energy consumption on the production side *before* wastes were made. This meant taking on the disposables market and challenging the legitimacy of built-in obsolescence. Reinforced by the escalating energy and oil crises of the 1970s, environmentalists argued that reusables and returnables made more sense than throwaways. With oil in short supply, the introduction of the first polyethylene terephthalate (PET) disposable soda bottle in 1975, made from petrochemicals (and invented by DuPont's Daniel C. Wyeth of the famous family of painters), amplified for some the hypocrisy and irrational wastefulness of the mass production system.[31]

Green Machine

What took the public so long to wise up to such destructive industrial practices? This delayed response resulted in part from the indefatigable efforts of manufacturers over the previous two decades to obscure the effects of so much waste and fend off challenges to its production.

In many ways industry's elaborate trash counteroffensive began with the earliest effort to restrict disposables in 1953. Coming not from nascent environmentalists but disgruntled dairy farmers, a measure passed by Vermont's legislature banned the sale of all throwaway bottles throughout the state. According to a newspaper report: "Farmers, who comprise nearly one-third of the House membership, say that bottles are sometimes thrown into hay mows and that there is need to prevent loss of cows from swallowing them in fodder."[32] Their livelihoods on the line, the husbandmen-politicians ratified the law to protect their animals from death by ingestion of stray glass containers.[33]

Within a few short months the packaging industry concocted a lavishly funded nonprofit called Keep America Beautiful (KAB), the first of many great greenwashing corporate fronts to come. KAB's founders were the powerful American Can Company and Owens-

Illinois Glass Company, inventors of the one-way can and bottle, respectively. They linked up with more than twenty other industry heavies, including Coca-Cola, the Dixie Cup Company, Richfield Oil Corporation (later Atlantic Richfield), and the National Association of Manufacturers, with whom KAB shared members, leaders and interests.[34] Still a major player today, KAB came out swinging, urgently funneling vast resources into a nationwide, media-savvy campaign to address the rising swells of trash through public education focused on individual bad habits and laws that steered clear of regulating industry.

Modeling themselves on beautification organizations like those later spawned by Lady Bird Johnson, KAB stealthily recruited local community and church groups, the public education system and government on all levels to join in their crusade.[35] In its first year, KAB printed instructive pamphlets and handbooks for schools in cooperation with the National Education Association to foster "lasting acceptance of good outdoor manners as a part of good citizenship."[36] Within its first few years KAB had statewide anti-litter campaigns "in progress or planned" in thirty-two states, membership ballooned to more than 70 million, and the group enjoyed the active support of four federal departments and fifty national public interest groups.[37] KAB was a fast success.

Legislation that cracked down on individuals who carelessly tossed their trash was welcomed by KAB. The organization supported stronger enforcement of local and state codes that, according to the *New York Times*, included "fines, with jail terms for repeaters, and uniformly strict application of the laws."[38] Such solutions suited the

KAB leadership, because they distracted the public from other options that might inconvenience industry, like production restrictions or forcing can, bottle and beverage makers to reinstate the vastly less profitable refillable container.

Their efforts got results. The Vermont law was defeated in 1957 with little press when the state legislature failed to renew the bottle ban. Now with no embargoes on packaging anywhere in the country, KAB and the interests it served controlled the floor.[39] The centerpiece of the organization's strategy was its great cultural invention: *litter*.[40] This category of debris existed before, but KAB masterfully transformed its political and cultural meaning to shift the terms of the garbage debate. KAB wanted to turn any stirrings of environmental awareness away from industry's massive and supertoxic destruction of the natural world, telescoping ecological disaster down to the eyesore of litter and singling out the real villain: the notorious "litterbug." Taking this tack, the group could defend disposability and obsolescence; the problem wasn't the rising levels of waste, they explained, it was all those heathens who failed to put their discards in the proper place.

KAB's approach was summed up in its production *Heritage of Splendor*, a 1963 "educational" film narrated by Ronald Reagan: "But while we're a responsible people with regard to our tangible resources like forests and minerals, how do we treat this important resource for recreation? We go away from home on a vacation and take a holiday from responsibility. We launch a *fallout* of litter. . . . Trash only becomes trash after it has first served a useful purpose. It becomes litter only after people thoughtlessly discard it."[41]

The key tactic of blaming individuals—as one American Can executive insisted, "Packages don't litter, people do"—obfuscated the real causes of mounting waste.[42] KAB paved the way in sowing confusion about the environmental impacts of mass production and consumption, today a favorite tool in the corporate greenwashing world. If the public believes that industry is responsibly handling natural resources, if they think production under a free market system is sustainable, and if average consumers accept that they are to blame when waste gets out of control—an Orwellian flipping of the script—then laws will not be enacted, government won't intervene, and production can continue on industry's terms. This sophisticated shaping of public opinion is a classic example of what the father of PR, Edward Bernays, called "the engineering of consent."[43]

KAB quickly apprehended that people's perceptions of waste could be influenced to better accommodate industry. This general dynamic is explained by philosopher Raymond Geuss: "Although reactions of avoidance and disgust seem to be rooted in basic facts of human biology and exist in all human societies, the particular form they take is culturally shaped and is acquired only through a long process of training."[44] KAB diminished industry's role in despoiling the earth, leaving it entirely out of their liturgy, while relentlessly hammering home the message of each person's responsibility for the destruction of nature, one wrapper at a time. And, with its slick PR machine, KAB successfully made all this Mad Tea Party logic appear utterly sane and normal.[45]

Such experience under its belt, KAB strode into the 1970s a seasoned anti-environmentalist, an old hand at turning the tables. On the second Earth Day in 1971, the group premiered the first of its now-iconic television advertisements starring the buckskin-clad longtime Hollywood actor Iron Eyes Cody. In previous decades, KAB got out in front of what the industry magazine *Modern Packaging* predicted would be "an awesome challenge" of reducing litter without restricting packagers, but now the group was practicing some serious jiujitsu.[46]

The haunting ad was seared into the guilty consciences of Americans young and old: After stoically canoeing through a wrapper- and can-strewn delta, past a silhouetted factory puffing smoke, Cody dragged his canoe onto a bank sprinkled with litter. Hiking to the edge of a freeway clogged with cars, the stereotyped Native American was abruptly hit on the moccasins with a fast-food bag tossed out the window by a freewheeling blond passenger. Cody then looked straight into the camera as he shed a single tear. The accompanying music was stirring, the voice-over solemn: "Some people have a deep, abiding respect for the natural beauty that was once this country. But some people don't. People start pollution. People can stop it."

The TV spot's inclusion of smokestacks and a traffic-choked highway marked a shift in KAB's message; maintaining its litter-is-the-root-of-all-evil position was getting more complex as the organization confronted the ecology movement. To maintain its legitimacy KAB was obliged to acknowledge issues like air and water pollution, against protests from its board of directors.[47] However, neither this commercial nor subsequent spots featuring Cody in similar dirty, tear-soaked scenarios departed from KAB's longtime message—the responsibility for pollution lay with the individual. Environmental devastation was the bitter consequence of each person's own selfish disregard for nature.

No Battle Too Small

The range of legislation debated across the country during the 1970s included outright prohibitions on one-ways, public oversight of packaging materials, mandatory deposits on disposables, and the implementation of municipal recycling. To combat this, the packaging industry and manufacturers doggedly pursued lobbying and legal work combined with media campaigns designed to sway public opinion to their will.[48]

These maneuvers were no doubt influenced by the brief but fierce 1959–60 struggle of the plastics industry to fight restrictions on polyethylene film when a rash of child deaths and adult suicides from suffocation induced newspapers and politicians to attempt to "Ban the Bags!"[49] In response, the indomitable Society of Plastics Industries (SPI) rallied its forces, launching a nationwide education campaign augmented by slick Batten, Barton, Durstine & Osborn ads promoting clear plastic bags as inherently good if used properly. During this battle, SPI also began working with its now-longtime counsel Jerome Heckman, who within a year's time traveled more than 40,000 miles defusing some sixty proposed legislative restrictions on polyethylene from federal, state, county and local bodies.[50] This pattern of opulently funding efforts to squelch regulatory measures on all levels of government would become the packaging industry *modus operandi*.

The next major battle over packaging came when Oregon state officials passed the first deposit law in the United States in 1972. Instead of outright forbidding throwaways, the measure imposed a five-cent refundable deposit on all beer and soda bottles and cans. The law also banned pull-tabs, effectively making all standard drink cans illegal.[51] Beverage and container producers lobbied hard to quash the bill's political support, arguing that the environmental effects would be insignificant, application of the law would be logistically impossible, and the whole affair would end in economic meltdown. KAB heavyweight American Can filed a lawsuit against the Oregon law that reached the state Supreme Court, but the bill survived.[52]

A year after its enactment, auditors conducted an assessment of Oregon's deposit law and found it was an indisputable success. Roadside litter was down 35 percent by volume; 385 million fewer beverage containers were consumed due to increased reuse and recycling; energy savings was sufficient to heat 50,000 Oregon homes; jobs increased; prices stabilized; and despite a brief initial slump in beer consumption, the market quickly recovered its normal growth rate. And the public liked the bill, giving it a 91 percent approval rating.[53]

A few months later, the Vermont state legislature enacted a similar law, which also survived industry's legal challenges. But beverage and packaging makers had other tricks up their sleeves to throw collection and reuse systems into disarray. As one Vermont resident explained:

> Our bottles bear so many types, sizes, colors, locations and positions of DEPOSIT notices, inevitably the public becomes confused. Most all of it is intentionally confusing. (All soda cans are marked on the bottoms, where you can't see the label; beer cans have the marking on top, very small and frequently right over the pull tab; many bottles have NO DEPOSIT NO RETURN *cast* into their sides, but have the *deposit* notice on the bottom.)[54]

Regardless, Vermont's law prevailed. And using Vermont's and Oregon's guidelines as models, the EPA conducted a study of projected five-year effects on litter, overall waste production, energy consumption, prices and jobs. The agency's findings revealed stellar results echoing those already recorded in Oregon.[55] At this point it was getting harder to believe industry's claims, like that from American Can executive William May, also a KAB leader, that container producers were "caught in the middle" by consumer pressure for new convenience packaging.[56]

In Minnesota, a tough legal battle broke out after the state legislature passed the 1973 Packaging Review Act. The act's Statement of Policy read: "Recycling of solid waste materials is one alternative for the conservation of material and energy resources, but it is also in the public interest to reduce the amount of materials requiring recycling or disposal." Explicitly favoring curbing consumption, the law also created public oversight power to restrict egregiously wasteful packaging. Container makers counterattacked with a PR campaign and a lawsuit charging that the measure unconstitutionally interefered with interstate commerce. The state's Supreme Court upheld the act, but gutted any means of implementation, leaving it to founder.[57]

Next came California. In 1974, the state's lawmakers brought a deposit bill up for debate, but this time KAB directly intervened. Active in the public relations and education spheres, KAB was not allowed to speak out on legislation because of its nonprofit status. Nevertheless, under intense pressure from the packaging interests within the group, KAB threw legal caution to the wind and sent its president, Roger Powers, to Sacramento to testify *against* the legislation. At the hearing Powers said, "We urge that you not take precipitous action on this piecemeal legislation, but that you examine the total picture. . . . We believe that such an investigation will prove that the proposed 'bottle bill' is not the answer."[58] California's law went down in defeat.

Eight years later, the citizens of California took another stab at regulatory legislation in the form of Proposition 11, which required a five-cent deposit on all beer and soda in bottles and cans. Industry's bitter attack on the bill reached new heights of mudslinging and mendacity, referred to in the press as "the year of the smear."

Just a week before the election, anti-Proposition 11 forces flooded California's airwaves with TV ads featuring interviews with "randrom shoppers" in Oregon who were harshly critical of their two-year-old deposit law. One man confessed to buying beer across state lines to bypass the burdensome measure. The people featured in the spots, however, turned out to be industry insiders who worked for businesses trying to stop California's pending legislation. The opposition also fabricated EPA reports to support their case, falsely claiming endorsements from politicians, and asserting that a bottle bill would hurt existing recycling programs, cause job losses, increase prices and create health problems (the latter charge being particularly bizarre).[59] Despite widespread public support for reducing trash levels to improve ecological systems at the start of the election, Prop. 11 was voted down.[60]

By 1976 every U.S. state legislature and numerous town, city and county councils—more than 1,200 in all—had proposed some form of restrictive packaging law.[61] According to an article from 1977 by environmental writer Peter Harnik, "Brewers, bottle and can makers, and soft drink companies sent lobbying teams to each legislature and made sure the bills either didn't come up for a vote or went down to defeat."[63] At a KAB board of directors meeting in the mid-1970s at New York's Biltmore Hotel, William May charged that deposit bill proponents were "Communists" and called for a no-holds-barred campaign against all upcoming bottle bill referenda.[64]

Of the three states, five counties and four cities that enacted packaging codes by 1976, most had been slammed with lawsuits by industry.[65] These legal offensives were often based on container and beverage companies' claims of unfair trade restrictions resulting from packaging laws. By the end of the decade only Maine, Michigan, Iowa, South Dakota, Connecticut and Delaware had managed to join Oregon and Vermont in adopting beverage container measures.[66] In most states only beer and soda containers were covered, while liquor, wine and juice packaging were exempted from the new laws due to harsh industry pressure.

On the federal level during the 1970s, manufacturers yet again deployed the guileful smokescreen of job losses and economic doom to head off nationwide packaging regulations—even after a presidential advisory committee recommended passage of a national bottle bill, calculating energy savings equivalent to nearly 5 million gallons of gasoline per day.[67] EPA and Federal Energy Administration (FEA) reports echoed those findings, indicating that a countrywide return to refillables would conserve almost 100,000 barrels of oil daily, increase employment and reduce the price of beverages by 30 percent.[68] That estimate made sense amid the decade's fiscal and energy crises. The FEA also found widespread public interest: 73 percent of Americans polled approved of a mandatory deposit law.[69]

In response to this momentum, Pepsi-Cola's then-president Donald Kendall pressured EPA chief Russell Train, an outspoken supporter of a federal deposit law, in a letter that read: "Your advocacy of a national beverage container deposit bill is directly counter to the President's war on inflation and his concern for unemployment. . . . Your position defies and denies the free will of the people expressed by their free choice of containers." Kendall sent carbon copies to Nixon and his chief of staff, Donald Rumsfeld.[70] By the 1980s, the beverage industry had purged most reusable packaging and the average consumer, regardless of what they really wanted, was left with the freedom of a narrowing choice of disposable, one-way containers and ever more massive mountains of garbage.

The Fox in the Henhouse

The captains of the industrial sector had to do some fancy footwork to beat packaging restrictions, and pitting labor against the environmental movement was their master maneuver. Using their perceptive PR machines and vast war chests, industry spread the word that what was good for the environment was bad for labor, a message that persists to this day (although the globalization movement has begun to debunk it). The common refrain—that reduced consumption necessarily meant declining production triggering inevitable job losses—veiled manufacturers' real concerns: lower profits. The claim of slumping employment was patently false, disproved by the jobs increase under the Oregon deposit law and by subsequent EPA findings.

The very companies engaged in industry consolidation that eliminated thousands of jobs over the preceding years argued that environmental laws would be the economic spoiler for the little guy, impinging on the high standard of living to which the average American had a natural-born right. This rhetorical feat was also remarkable because it defied a key reality: labor was lost when commodities got destroyed through landfilling and incineration. As Karl Marx wrote, "The worker puts his life into the object; but now his life no longer belongs to him but to the object."[71] If this is true, then in our current system, the worker's life belongs not only to the commodity he or she makes, but also increasingly to the garbage graveyard. When manufactured goods are trashed, so too is the labor that went into making them.

In this regard, garbage is not just nature, but human labor thrown away in the interest of the circulation of commodities and the extraction of profit. Or, if we were to modify Marx's famous formulation, the letter that drops out of M-C-M' is G for garbage.[72] Labor researcher John Marshall explains this connection between labor, profits and garbage:

> The important thing to remember about landfills is that they're not just an unfortunate byproduct of capitalism; they actually represent the success of capitalism. The extent to which profit has been achieved is the extent to which we have more garbage to put in the landfill.
>
> But we also need to think about the amount of labor that went into producing these products; the time that was spent making these products that were specifically designed to wear out faster than they needed to. What else could we have done with that time? What other productive, scientific, creative pursuits could we have invested that time in? What kind of society could we have created alternatively?[73]

The calculated undermining of the potentially powerful alliance between labor and the ecology movement served capital well. Generating confusion about the real issues allowed business to derail popular regulatory bills, extending to industry largely unfettered access to natural resources for expanding mass production. The industry also managed to keep deposit laws to a minimum. Today only eleven states have mandatory deposits on cans and bottles.

While fighting on legislative and legal fronts, the packaging and related industries have used the coercive powers of PR to a chilling extent. "The 'engineering of consent' implies the use of all the mechanics of persuasion and communication to bend others, either with their will or against their will, to some prearranged conclusion, whether or not their reaching that conclusion is in the public interest," wrote PR don W. Howard Chase in explaining Edward L. Bernays's core idea.[74] Industrial production and resource extraction have thus taken shape outside the realm of public participation and are based on increasing profits and expanding markets, not human and environmental health solidly rooted in democracy.

Stored bales of plastic and metal at a recycling center. (Still from the documentary *Gone Tomorrow*)

7

Recycling: The Politics of Containment

The national average price for dumping trash in landfills remained consistent between the 1950s and the early 1980s, but between 1984 and 1988 the cost suddenly more than doubled.[1] In cities like Minneapolis, rates spiked sixfold from $5 to $30 per ton.[2] A major mechanism behind this surge in fees was new enforcement of the ten-year-old Resource Conservation and Recovery Act (RCRA), which contained a provision—referred to as Subtitle D—that required safety standards for land disposal sites and was the first federal effort to regulate waste facilities.

The vast majority of landfills from the postwar era through the late 1980s accepted a wide range of hazardous materials and lacked control systems for poisonous leachate and noxious gas. Even if landfills utilized the methods Jean Vincenz pioneered earlier in the century, over time sanitary fills poisoned soil, contaminated groundwater and polluted the air. By the late 1980s, so many former fills has been declared toxic that they comprised half of all Superfund sites—among them Vincenz's Fresno sanitary landfill.[3] Finally implemented thanks to shifting political will, Subtitle D aimed to force aging facilities to clean up or shut down.[4]

Enforcement of Subtitle D had a dramatic effect. At that time, 90 percent of U.S. wastes were disposed of on land and an estimated 94 percent of those sites were not up to standards.[5] As a result, U.S. landfills were shuttered by the hundreds, their numbers plummeting by two-thirds during the 1980s.[6] At the same time, garbage output was exploding; between 1960 and 1980 the amount of solid waste in the United States quadrupled.[7]

Also putting pressure on landfill operations was the power that new suburban jurisdictions now exercised over the use of their land. In previous decades, municipalities that needed additional disposal space could just head to the edge of town and start dumping. Now urban officials were faced with suburban local governments that blocked the construction of nearby landfills, a move that drove waste treatment farther afield. Siting landfills at great distances from population centers pushed rubbish transportation costs too high for many municipalities and waste handling firms. This meant that finding appropriate places to build new fills was difficult at best, a reality that compounded the disposal crisis already under way. Municipalities had to find other solutions, and find them fast.

The pro-waste bloc in industry and government, and, perhaps surprisingly, some of the high-profile national environmental organizations, initially advocated the filthy standby, incineration. Beginning its steady shift toward policies that favored industry, the EPA also supported incineration as the answer to the disposal crunch. The justification among old-line environmental groups and the EPA was that incineration was an improvement on open burning and unmonitored dumping. Into the 1980s, they argued that burning refuse was a safe, appropriate and "proven" disposal method. And since turning trash into ashes reduced waste volume, incineration would ease pressure on shrinking landfill space. Bolstered by a program of subsidies for incineration through the Department of Energy under President Jimmy Carter, burning seemed destined to take over the garbage disposal market.

However, to the waste industry's surprise, in the late 1980s and early 1990s grassroots resistance to smoke-belching incineration stopped the proliferation of incineration plants. Across the country, diverse neighborhood groups concerned with the crosso-over of ecological and social justice issues began forcing municipalities to abandon rubbish cremation and pushed forward a new era in popular environmentalism dominated by recycling. By the late 1980s, there were more than 5,000 municipal recycling programs in the United States, up from just 10 in 1975.[8]

Under mounting public pressure, waste companies, their allies in public office and many in the manufacturing sector reluctantly accepted some forms of recycling. Given the range of possibilities, reprocessing wasn't the worst-case scenario for industry. After all, recycling would cause the least disruption to existing production processes. More drastic measures—like reducing consumption (through output regulations) or mandating reuse—were far more threatening. Recycling continued to treat wastes *after* they were created. And recycling was a technological fix that required its own manufacturing processes, ensuring continued production.

Wildly popular with the public, recycling is certainly more environmentally sound than incineration or landfills, but it has a number of limitations. Not all those carefully sorted discards actually get remade into new products; many separated materials are just hauled to the incinerator or landfill for disposal anyway, due primarily to lack of markets. As of 2000, U.S. recycling rates were surprisingly low: 54 percent for aluminum, 26 percent for glass, 40 percent for paper, and a paltry 5 percent for plastics.[9]

While offering real and substantial benefits, recycling has also functioned in cultural terms to divert public attention away from stronger reforms. The promise of recycling further normalized growing consumption, telegraphing to the public that even in the act of discarding one could be environmentally responsible; tossing the empty bottle, the once-used piece of paper, or the cereal box into a special bin took the guilt out of so much wasting. Recycling also became a vehicle for manufacturing to reinvent itself as an environmentally responsible caretaker of the planet, unleashing a new phase in corporate greenwashing.

Despite its shortcomings, however, recycling has proven to be a more efficient use of resources than burning and burying. The practice has also fostered a growing public consciousness about the environment and helped maintain popular awareness of the link between industrial production and ecological and human health.

Burn, Baby, Burn

In addition to Subtitle D, the federal measure creating minimum safety standards at land disposal sites, a dramatic shift in solid waste policy at the EPA and massive federal subsidies underpinned the re-emergence of incineration.

In 1978, as a response to the energy crisis, the Carter administration adopted the Public Utilities Regulatory Policies Act (PURPA). In addition to other provisions, the act offered a plethora of subsidies and breaks to incinerator firms under the guise of creating "energy independence." Because new trash-burning plants promised to turn the energy released in combustion into electricity, they qualified for a wide range of subsidies under PURPA. But again, this policy steered clear of real changes that could reduce energy consumption from the start.

Much of incineration's newfound popularity rested on semantics. Updated incinerators (many constructed by companies rerouting their capital and expertise away from building controversial and increasingly unpopular nuclear power plants) continued the late-nineteenth-century practice of capturing energy given off in combustion for use as a power source. Proponents called this process "waste-to-energy" or "resource recovery." While these plants did capture heat and steam generated during the incineration process, the amount of energy created was a small fraction of what could be conserved if less was wasted in the first place. Twice as much energy would be saved for every ton of trash not produced (through reusing materials and reducing consumption) than power captured by torching that same amount of garbage.[10]

In contrast to its early days, by the late 1970s the EPA was filled with industry-friendly bureaucrats uncritical of prowaste policies. The EPA worked with the Department of Energy (DOE) to encourage incineration, formally designated as EPA's disposal technology of choice. Most notably, under PURPA, the EPA and DOE made regulatory adjustments and worked with industry to promote incinerator construction by creating grants, below-market-rate loans, loan guarantees, arbitrage and municipal bonding rules, and price supports; guaranteeing resale of electricity generated at incinerators; and reclassifying ash as a nonhazardous material.[11]

In this climate, incinerator construction positively flourished. In 1980 the number of trash-burning plants online, proposed or under construction was about sixty. Five years later that number more than tripled and, a couple of years after that, more than three hundred were fired up or in the pipeline. The EPA predicted that incineration was "expected to grow at an astonishing rate . . . significantly faster than the growth rate for municipal refuse generation."[12]

Down in Flames

No sooner had the new plants begun stoking their furnaces than public outrage over mass burning began to smolder. In a world incorporating environmentalism into its popular consciousness, garbage incinerators triggered a new and more sharply focused brand of community protest. The EPA had already pinpointed, in the late 1970s, "public opposition to new facilities" like incinerators as "the major problem" confronting efforts to update and expand solid waste management.[13] For those living near proposed sites, the obvious increases in pollution from added traffic, diesel exhaust, and incinerator ash and smoke were too much to ignore. And evidence was mounting that a group of highly toxic chemicals known as dioxins were spewing from furnaces burning ordinary residential garbage. Armed with an increasing knowledge of the health dangers from cremating wastes, local residents rose up in neighborhoods across the United States in unexpectedly diverse coalitions to oppose incineration.

When New York City proposed building eight incinerators in the early 1980s, it was among the first municipalities to encounter this formidable resistance. The initial refuse crematory was to be sited in working-class Williamsburg's Brooklyn Navy Yard. Initially, locals resisted the plant because of greater truck traffic. But after incinerators like that in Hempstead, Long Island, were found to emit unexpectedly dangerous levels of dioxins, community members understood the plant's more hidden and more serious risks.[14]

Overcoming longstanding racial divisions, Williamsburg's Hasidim and Latinos joined forces with African Americans from nearby Fort Green to form the Community Alliance for the Environment to stop the Navy Yard incinerator. Educating themselves in the dry science of incineration and air pollution, with the technical assistance of Barry Commoner's Center for the Biology of Natural Systems, (CBNS) and support from the New York Public Interest Research Group, the Natural Resources Defense Council and other environmental organizations, Williamsburg's residents waged a long and dramatic battle for safe emission levels from the proposed trash burner.[15]

During the 1980s more was discovered about dioxins. Most famous as a hazardous ingredient in Agent Orange, dioxins are the unintended by-products of industrial production, but they also result from incinerating everyday trash. While researching the Navy Yard incinerator, Commoner's CBNS found what other scientists at the time were discovering: materials containing chlorine compounds like plastics, table salt and bleached paper that were burned together with organic matter that contained carbons like wood formed new bonds that created a net *increase* in dioxins. Instead of destroying the poisonous group of chemicals, incineration actually produced more dioxins. By the late 1980s some of those dioxins were revealed to be among the most toxic molecules known. Dioxins are carcinogenic, reduce fertility, affect fetal development, cause the skin condition "chloracne" and compromise the immune system, radically escalating susceptibility to infectious diseases. Dioxins are considered highly toxic because they accumulate in the human body over time, they have a wide array of adverse health effects, and they can cause harm at very low exposure levels.[16]

The incinerator industry addressed these growing concerns by introducing filters to cut dioxin releases. This meant a "scrubber," which would cool and add lime to the flue gases, followed by a fabric filter that could potentially capture 90 to 95 percent of dioxins. But, according to Commoner, the promise of eradicating dioxins was illusory: "While such controls correspondingly reduce the amount of dioxin emitted from the stack, they do so not by destroying it, but by transferring it to the fly ash captured by the scrubber and filter. Instead of entering the environment through the air, now most of the dioxin enters the environment when the fly ash is removed from the control device and disposed of." And, once buried, dioxins can easily seep into groundwater and contaminate soil. Additionally, during the 1980s heavy metals like lead, cadmium and mercury were found in incinerator waste ash and identified as toxic risks.[17]

In the end, New York's Department of Sanitation was unable to prove that dioxins could be sufficiently controlled even using the most cutting-edge technology. Several years of focused community pressure successfully blocked the Navy Yard incinerator and, in the mid-1990s, funding for the project was cut. Other skirmishes erupted in Gotham's various boroughs as residents organized against the city's plans to build the remaining trash crematories, and none of the eight on New York's drawing board were ever constructed, thanks to organized and committed community opposition.[18] Meanwhile, similar battles were raging on the West Coast.

In 1985, the California Waste Management Board, the state's main refuse handling advisory body, warned that California would run out of landfill space by 1997 and recommended increased use of incineration as the solution. A minimal role was given to recycling, while reduction and reuse strategies were ruled out entirely, due to industry pressure to leave production levels untouched.[19]

By the mid-1980s California had plans for thirty-four incinerators with three in Los Angeles alone. Known as the Los Angeles City Energy Recovery Project (LANCER), the trio of trash-burning facilities enjoyed wide support from local politicians, including mayor Tom Bradley, and they were massive; together the furnaces could burn up to 70 percent of L.A.'s waste. The first of the three LANCER incinerators was slated to be built in the poor, working-class Latino and African American neighborhood of South Central.

Not believing the Bureau of Sanitation's assurances of safety (the agency had pledged that the facility would be so clean that local families could hold wedding ceremonies on the lawn), a group of residents organized into what would become an unstoppable force—Concerned Citizens of South Central Los Angeles (CCSCLA). As in Williamsburg, these residents were worried about the health effects of dioxin and toxic metals released from the plant.

In rallying opposition, CCSCLA built coalitions with an impressive array of community and professional groups. The group worked closely with other citizens', homeowners', and environmental "slow growth" organizations from a range of neighborhoods across Los Angeles, including the wealthy West Side, also due to receive one of the three planned incinerators. Before long, CCSCLA had won over several city council members, UCLA urban planning students, and local and national environmental groups, including Greenpeace.[20]

Trash-burning proponents never expected the diversity and resulting strength of the opposition and were therefore caught completely off guard. Ultimately the mayor and many other local officials switched sides, and in 1987 canceled the entire project. In direct response to the anti-incineration movement, several L.A. City Council members began promoting municipal recycling.[21]

The undoing of high-profile projects like LANCER dealt a mortal blow to incineration. Since it involved some of the field's most influential companies overseeing state-of-the-art operations, LANCER had become a new standard-bearer; when it was defeated, the incineration industry was rocked to its core.[22]

Forty planned trash crematories were stopped between 1985 and 1989 in cities such as San Diego, Seattle, Boston, Kansas City and

Philadelphia, with the latter going so far as to adopt an explicitly anti-incineration solid waste management plan that mandated recycling.[23] An investment firm that previously promoted bond issues to finance incinerators warned that "public opposition will remain the toughest hurdle for this industry over the next several years. Concerns over dioxin emissions and ash residues high in heavy metal concentrations will continue to play a major role in this industry's development."[24] The *Wall Street Journal* reported problems for the incineration industry in 1988: "More than $3 billion in projects have been scrapped in the last 18 months and new orders have slowed to a trickle."[25] By 1992, only 10 percent of U.S. waste was sent to incinerators, compared with 80 percent still shuttled to bulging and substandard landfills.[26]

The incineration industry labeled this effective community opposition NIMBY (the acronym for "not in my backyard"). However, instead of shortsightedly advocating the siting of trash-burning facilities somewhere else, growing numbers of antiburn activists opposed incineration everywhere. Many of these grassroots organizations crystallized into local, regional and nationwide networks with a broader vision. By linking toxicity with escalating consumer wastes and social and economic injustice, this new incarnation of environmentalism expanded to include issues of class, race and labor.

Hundreds of communities had been fighting their own environmental battles for years, but it wasn't until the late 1980s and early 1990s that a series of events galvanized these activists into a self-aware national (and then global) movement. A 1987 report by the United Church of Christ's Commission for Racial Justice, titled

"Toxic Wastes and Race in the United States," documented that a geographical area's racial composition was the single most reliable factor in predicting the location of waste disposal sites. This report confirmed a larger pattern of discrimination and was circulated widely among neighborhood organizations around the country. The next milestone was the 1991 National People of Color Environmental Leadership Summit where hundreds of grassroots groups joined forces. Richard Moore, who for years had fought toxic dumping in his hometown of Albuquerque, New Mexico, noted the far-reaching aim of the burgeoning movement: "There was a tendency toward 'Not in My Backyard.' What we said very clearly at the People of Color Summit was, 'Not in Anybody's Backyard.'"[27]

The Carter administration had done a lot to promote incineration, but once the cardigan-wearing president was no longer in charge of the White House thermostat, Ronald Reagan dealt a perhaps unexpected blow to incineration advocates. He cut the industry off from government subsidies, believing that, like any good business operating in the free market, trash burning companies should fend for themselves. Forced to compete on the open market, and faced with the formidable opposition of groups fighting environmental racism, incineration was in serious trouble. Consequently, burning garbage failed to gain any significant share of the disposal market and has receded, at least for a while, into the background.

Recycling as Panacea

With incineration blocked and measures like mandatory deposit laws and source reduction getting smothered by industry, recycling underwent a renaissance in the 1980s and early 1990s. Across the country, schoolchildren put down their glue and macaroni to learn the virtues of "closing the recycling loop." Suburban housewives made room in their kitchens for extra bins to separate paper from cans and bottles. Community activists and neighborhood groups were revitalized, sparking renewed interest in recycling from politicians as municipalities began experimenting with and implementing recycling programs.

In the 1980s curbside recycling systems were adopted—many of them mandatory—in Connecticut, New Jersey, New York, Rhode Island and Maryland, while dozens of American cities and counties passed their own measures.[28] One of the earliest curbside trash collection systems was in Fitchburg, Wisconsin, a suburb of Madison. Fitchburg's city council fired the country's largest trash company, Waste Management Inc., and brought in another firm to devise a more comprehensive recycling program. Fitchburg's new plan was based on an innovative system in San Jose, California, which used brightly colored plastic bins distributed to residents for storing separated recycling—not unlike the nineteenth-century sanitarian Colonel Waring's "source separation." Fitchburg's mandatory program saw a 66 percent participation rate in the early months of operations, confirming recycling's popularity and its promise to make incineration unnecessary while diverting materials from bulging landfills.[29]

East Hampton, New York, adopted an intensive recycling regimen in 1989 as its landfill was reaching capacity. Begun as a pilot program designed by Commoner's CBNS, the scheme still sounds cutting-edge. Each participating resident received four color-coded tubs, one for compostables (food and soiled paper), another for cans and bottles, the third for unsoiled paper and cardboard, and the last for nonrecyclables. The trial showed that only 13 percent of East Hampton's wastes needed to be disposed of—the remaining 87 percent were recyclable. East Hampton's recycling program could handle as much waste as a proposed incinerator, but would cost 35 percent less. The Town Board quickly approved the plan.[30]

By 1993, almost a quarter-century after the first Earth Day, the EPA reported that domestic recycling had tripled by weight, from 7 percent to almost 22 percent.[31] More cities were passing voluntary and mandatory recycling measures, and growing numbers of Americans were welcoming recycling into their daily routines.

Damage Control

With all this recycling, one might think there had been a green conversion on the highest levels of industry. While beverage producers began feverishly stamping their bottles and cans with the "chasing arrows" recycling symbol and proudly advertising more enlightened, eco-friendly ways, countless executives were actually squirming in their seats. The decades-old consumer joyride had hit a snag. Industry was forced to make unwanted changes due to the very real crisis over where to put the trash, and a public more sensitive to environmental degradation.

These tensions peaked in one of the decade's most influential incursions against trash, the McToxics Campaign. Coordinated by the Citizens Clearinghouse for Hazardous Waste, a grassroots group formed by victims of the Love Canal disaster, the effort targeted McDonald's Styrofoam "clamshell" packaging as a needless and toxic waste. By the late 1980s, according to an EPA study, more than 99 percent of all plastic containers were discarded after only a single use. Americans were throwing away 10 million tons of plastic each year, amounting to a bulky 25 percent of all waste by volume.[32]

Participants in the national campaign led pickets, boycotts and local efforts to adopt bans on Styrofoam fast-food packaging. A nationwide "send-it-back" mail-in of used clamshells to McDonald's executives returned mounds of refuse to the country's single largest consumer of foamed polystyrene containers.[33] Coinciding with ongoing headlines about CFCs chewing away the ozone layer, the McToxics Campaign and another effort by Friends of the Earth lent momentum to regulatory laws. Maine and Vermont, as well as Berkeley, California, and Suffolk County in New York initiated efforts to ban fast-food foam packaging, particularly those items made with CFCs. By 1988, twenty-one other states were debating similar restrictions.[34]

After a fierce three-year battle during which McDonald's proposed in-store mini-incinerators to burn used packaging on site and described Styrofoam as "basically air" that was benign in landfills because it "aerated the soil," the fast-food giant gave up using the stuff in 1990.[35] A domino effect set in as other fast food restaurants across the country ditched foamed plastic containers. But perhaps more significantly, the highly visible and popular campaign signaled a shift in the way Americans regarded their influence over production. By fighting the use of packaging made with CFCs, the campaign highlighted that not only were people no longer willing to faithfully accept waste and toxicity, they were capable of changing the production process itself.

Business interests were on the defensive. After all, the stakes couldn't be higher. With landfill space shrinking, new incinerators ruled out, water dumping long ago outlawed, and the public becoming more environmentally aware by the hour, the solutions to the garbage crisis were narrowing. Looking forward, manufacturers must have perceived the range of options as truly horrifying: bans on certain materials and industrial processes; production controls; minimum standards for product durability; higher prices for resource extraction.

An oasis in this desert of regulation, recycling offered industry what other options lacked. Unlike reduction and reuse, remanufacturing single-use materials into new containers and packages would not impinge on consumption levels and would have the least impact on established manufacturing processes. That meant producers could keep making and selling disposable commodities in much the same way that they had been doing for decades. As American Can Company executive William May had put it a few years earlier when discussing the potential garbage tidal wave from so much disposable packaging: "We must comprehend, as a nation that the solutions also lie, to a very large degree, in technology. . . . We oppose any reduction in productivity."[36] In the industry's eyes, recycling was the lesser evil.

As recycling spread in the 1980s, promoted by some in industry and some environmentalists, manufacturers were not on altogether unfamiliar turf. Not only had the packaging industry itself created the recycling symbol—the Container Corporation of America commissioned the design a few months after the first Earth Day to advertise its reprocessed products and left the logo in the public domain for others to adopt—manufacturers also pioneered modern recycling procedures.[37] In the 1970s some producers conducted their own experiments in reprocessing, most likely to establish some measure of control over future recycling standards and to get out in front of the ecology movement in shaping popular perceptions of the still largely undefined practice called "recycling."

One of the earliest such trials was a facility opened in 1973 in New Orleans by the National Center for Resource Recovery (NCRR). The nonprofit research organization was started in 1969 by KAB and received funding from container-related companies.[38] NCRR signed a contract with the city to process its recycling, but on the condition that a preset volume was maintained, an agreement that precluded any reduction in discards.[39] A more elaborate pilot project was unveiled in Milwaukee in 1976 by another major KAB player, the American Can Company. The cutting-edge facility was dubbed "Americology." By that time American Can had become a huge conglomerate and major producer of disposables with a range of can factories, paper mills and packaging and food service materials plants.

Through a contract with the city, municipal collectors delivered mixed garbage to the floor of Americology's "sparkling new plant" where "inspectors" and high-tech machines separated the discards into "recoverable components." As detailed by former American Can executive Judd Alexander in his book *In Defense of Garbage*,

Americology's technologies were strikingly similar to those used today. They included "air classifiers," which used bursts of air to segregate lighter from heavier discards; an "eddy current separator" that sorted aluminum by electrical impulse; and a series of vibrating screens that sifted out glass, ceramics, stones, soil, coins, "and an occasional bullet or shell casing." American Can had visionary plans for the plant that included "a flotation system to separate the glass which would then be classified by color with an optical reader."[40]

Despite all this automation and an abundant market for recyclable metal, within a short time Americology closed due to economic troubles. The plant's financial problems stemmed from a failed scheme to sell processed trash as a coal substitute to a local power plant. The power company quickly realized that coal was much more efficient and cleaner to burn than garbage. Americology's building was later reopened as a transfer station where Milwaukee's wastes were processed for landfilling.[41]

During the peak of 1970s environmentalism, many U.S. companies, fearing the tide had turned against them, lobbied state, local and national politicians for policies that favored recycling. They did this as a bulwark against adoption of more rigorous measures mandating reduction and reuse. Writing at the time, Environmental Action's Patricia Taylor observed that, in contrast to efforts that would curtail trash production, recycling had "a frightening potential for institutionalizing waste generation."[42] Her assessment was right; recycling did allow for maintaining and further ingraining high levels of production, consumption and waste, a key selling point for embattled manufacturers. Around the same time, NCRR aggressively courted officials at the highest levels, even sending representatives to meet with Donald Rumsfeld, then Nixon's chief of staff.[43] So in the 1980s, when recycling became a necessity, the industrial sector had already laid important political groundwork.

Green All Over

Stiffer regulations on production narrowly averted, manufacturers of all kinds began to comprehend the valuable PR and marketing opportunities that could spring from recycling and green branding. With greater numbers of Americans identifying as environmentalists, why shouldn't big business do the same? As it turned out, recycling helped industry in two ways: it distracted the public from making more radical demands while offering a means to connect to a new consumer base. This played out most strikingly in the plastics packaging sector.

One early resin greenwashing episode occurred in 1988, when the cunning industry group the Society of Plastics Industries (SPI) took up the chasing arrow symbol that was at the time embraced by the ecology movement. The decades-old SPI altered the image slightly by inserting numerals in its center, which assigned various plastics grades 1 through 9. The SPI then promoted this to state governments as a "coding system" that could be used in lieu of restrictions like bans, deposit laws and mandatory recycling standards. The industry-backed Council on Plastics and Packaging in the Environment explained in a 1988 newsletter that some state legislatures adopting the coding system had done so "as an alternative to more stringent legislation." At the time almost $1 of every $10 Americans spent for food and beverages paid for packaging; with such a tremendous market, the fast-growing plastics industry was doing all it could to shut out regulation.[44]

Once states adopted the plastics codes, resin packaging was stamped according to its grade, but, significantly, it was also embossed with the recycling symbol. Indicating nothing specific about recycling, plastic packaging bearing the triangular symbol misleadingly telegraphed to the voting consumer that these containers were recyclable and perhaps had even been manufactured with reprocessed materials themselves. But often neither was the case.

While the code numbers did indicate various types of plastics in the broadest sense, there were (and are) so many critical variations within those categories that this system's efficacy was (and remains) questionable. And, crucially, these grades did not create a sorting system truly useful for producers, since resin makers must categorize discarded plastics based on how they're made, regardless of their code number.[45] As a result, not only was SPI's grading scheme a hollow substitute for real, meaningful programs, but by the early 1990s these stamps were criticized by some recycling centers as undermining environmental goals, creating public confusion over what was actually recyclable, and driving up costs for local facilities that were left to handle these wastes.[46]

In 1993, the Environmental Defense Fund slammed plastics makers over their measly recycling numbers, charging that the rate of recycling "did not even come close to keeping up with increased production of virgin plastic. . . ."[47] Just four years later, new plastics production outpaced resin recycling by nearly 5 to 1.[48] Thanks to the polymer trade's astute green marketing, however, most Americans believed otherwise.

In its next phase of corporate greenwashing, again with the help of the powerful SPI, the plastics industry established the American Plastics Council (APC). An industry-backed organization like KAB (with whom the group shared members and leaders), the APC conducted education and PR campaigns touting the virtues of recycling and, of course, the benefits of plastics. With much fanfare, the APC announced a 25 percent recycling goal for plastics, spending $40 million in 1992–96 to publicize the wholesome goodness of resins. While buffing its green veneer, the APC operated behind the scenes to prevent legal restrictions on resin makers. As *Plastics Engineering* magazine's senior editor put it, the APC worked to "assure a measure of restraint and reason in the drafting of packaging legislation."[49] During the mid-1990s the "pro-recycling" APC and SPI actively opposed the passage of some 180 regulatory and legislative proposals in thirty-two states. And, with no legally binding measures in place, in 1996 the APC quietly abandoned its 25 percent recycling target.[50]

All this greenwashing was possible through massive spending on public relations. And, as Marx pointed out:

> I am ugly, but I can buy for myself the most *beautiful* of women. Therefore I am not *ugly*, for the effect of ugliness—its deterrent power—is nullified by money. . . . I am bad, dishonest, unscrupulous, stupid; but money is honoured, and therefore so is its possessor. Money is the supreme good, therefore its possessor is good.[51]

If money can make the bad good, and the stupid smart, then it can make the ecologically destructive green.

As the 1990s wore on, the chasing arrows and printed appeals to "please recycle" adorned every type of bottle, jar, can, envelope and wrapper, as if they were endorsed by Mother Nature herself. The social and political impact of all this pro-recycling PR was much like that of anti-littering efforts in previous decades. Regardless of industry's actions, the rhetoric of recycling targeted individual behavior as the key to the garbage problem, steering public debate away from regulations on production.

The Realities of Recycling

As recycling fever spread, more Americans believed they were helping reverse environmental decline, but the truth was not so simple. It was apparent by the end of the 1980s that remanufacturing procedures could be inefficient, polluting and expensive. While recycling rates had grown considerably over the previous twenty years, in the mid-1990s they began to stagnate while wasting rates continued to climb.[52] In large part this was because recycling did not minimize the creation of discards. Instead, this back-end refuse management strategy left wasteful mass production and consumption unaltered and even encouraged. People started believing that their trash was now benign. Today it's likely that more Americans recycle than vote—yet greater amounts of rubbish are going to landfills and incinerators than ever before.[53] Far better than directly sending discards to be burned or buried, recycling nevertheless has serious flaws.

Just because materials are hauled away in a recycling truck doesn't mean that they actually get reprocessed. Almost half of discarded newspapers and office paper is buried or burned, while two-thirds of glass containers and plastic soda and milk bottles are trashed instead of recycled.[54] If wastes are remade, they immediately hit the hurdle of "downcycling." All substances, except for some metals, lose their molecular integrity during reprocessing, eventually rendering them unusable. For example, each time it gets recycled, paper's long fibers break, becoming shorter and less able to hold together. Similarly, remelted glass loses its workability and durability with successive reprocessing. The least recyclable of packaging materials, plastic, loses its infrangible flexibility when made molten again. Also, synthetics are highly sensitive to contamination, so even trace levels of a different type of resin, say a stray laundry detergent cap mixed with milk jugs, can mean that the whole lot gets trashed. (A homogenous type of plastic, say PET with no contaminants, downcycles less rapidly.) Moreover, a huge proportion of virgin resin must be mixed in to reinforce the weakened plastic to create a useful substance. Because of downcycling, recycling has inherent technical limits.[55]

Also, many "recyclables" are remanufactured only once. The most common type of plastic reprocessing in the United States uses recovered resins to produce new commodities that are usually not recyclable themselves like fleece, car bumpers and synthetic lumber. After they wear out, these recycled products must be burned or buried. So, while it might sound neat that used water bottles get remade into sweatshirts, these items' lifespan is finite—thus failing to form a "closed loop" of recycling.[56]

Energy-intensive remanufacturing distinguishes recycling from other waste-cutting options like reuse and reducing consumption.[57] Making a barely used commodity anew drains vastly more natural resources compared to, for example, washing, sterilizing, and refilling a reusable bottle, as is commonly done in Latin American, Asian and European countries today. Additional energy is also needed for managing and transporting materials during the recycling process. At least 20 to 30 percent of U.S. plastic recyclables are exported to other countries, mostly in Asia.[58] That means a huge portion of once-used soda and milk bottles get shipped all the way across the Pacific so they can be sorted, remelted and turned into products that themselves might not be recyclable. Furthermore, a Greenpeace seven-country survey found that up to 50 percent of plastics discards shipped overseas were so contaminated with incompatible materials that they could not be recycled and were instead dumped, often in unlined, unmanaged sites.[59]

Hazardous contamination can also result from recycling. Used paper de-inking mills are notorious for releasing toxics into the environment. Fort James Corp., a major paper maker, operates a recycling facility in Green Bay, Wisconsin, that is the second largest individual polluter in the state. Aluminum reprocessing is also known to wreak havoc on natural systems. A malignant by-product of the resmelting process, aluminum dross—a chemically active waste exempt from federal regulations—is frequently dumped on open land, left to contaminate soil and seep into groundwater.[60]

If materials make it through the reprocessing phase, they face yet another hitch: the market. Selling recycled feedstock is often hampered by unpredictable demand and prices, exacerbated by lower costs for virgin resources. Using recycled materials is vastly more efficient, but it appears more expensive thanks to a distorted economic picture. Recycling must compete with extractive industries that have built their infrastructures with decades of state support and still receive substantial direct and indirect aid from federal and state tax breaks, energy subsidies, and road building, adding up to billions of dollars annually.[61] And with no meaningful recycled content laws on the books, producers most often don't use reincarnated raw materials in their goods. If virgin feedstock costs less on the market, as is often the case, manufacturers are under no obligation to buy recycled materials.[62] A further measure of industry subsidies through cost externalization is taxpayer-funded waste management for all those trashed commodities. The United States spends over $43 billion annually to collect and dispose of municipal wastes, 76 percent of which are manufactured goods.[63]

Creating other ongoing complications, recycling is constantly under threat because it also must outperform disposal. Again, in this realm the playing field is not level. Recycling programs are expected to pay for themselves or even turn a profit, while, held to a very different standard, solid waste departments are fully funded no matter what. If recycling doesn't bring in revenue it often risks getting cut from both private waste companies and municipal governments.

Ironically, many recycling operations are overseen by waste handling firms that earn exponentially more by trashing rather than reprocessing. One report estimated the Waste Management Inc.'s profit margin on landfills was ten times that of recycling.[64] After consulting with waste industry officers, Morgan Stanley Dean Witter concluded that "recycling has long been the enemy of the solid waste industry, stealing volumes otherwise headed for landfills."[65]

Forced to compete under these circumstances, discard dealers often resort to wasting even if they don't want to. David Kirkpatrick, owner of a Durham, North Carolina, recycling consulting firm, explained this double bind: "It's a weird system, because if you do a better job with recycling and reducing your garbage, then you're also reducing your revenues."[66]

Faced with budget crises, Washington, D.C., halted municipal recycling several times in the mid-1990s, as did New York City for two years starting in 2001. In both cases, local officials argued that they had to suspend recycling to save money. (And in both cases, tremendous public pressure to reinstate recycling brought the programs back.) But if recycling received a fraction of the state assistance that manufacturing and refuse disposal have been given over the last century, the facts on the ground would be flipped. The oft-repeated and bizarre argument that recycling is inadequate because it can't compete on the open market is dubious at best.

Recycling Reconsidered

Despite its many drawbacks, recycling has real benefits. It's far better to reprocess materials than to send them to be burned or buried. As the Environmental Defense Fund has pointed out, "The energy conserved through recycling is about five times as valuable as the average cost of disposing of trash in landfills in the U.S."[67] Recycling is also far more efficient than manufacturing with virgin materials. According to the GrassRoots Recycling Network, aluminum cans made from reprocessed material require 96 percent less energy to manufacture, and PET (soda) and HDPE (milk) containers made from scratch use four to eight times more power than if they were manufactured with reincarnated resin. Other environmental benefits from recycling are equally dramatic. Reconstituted pulp requires 58 percent less water and creates 74 percent less air pollution than making paper with wood pulp from freshly cut trees.[68]

Expanded use of recycled materials in industrial processes would have far-reaching effects, since most refuse is created during production. Municipal waste, including discards from households, local businesses and institutions like schools, accounts for less than one in every seventy tons of castoffs. The rest—the great majority—is generated by industrial processes like manufacturing, mining, agriculture, and oil and gas exploration, so cutting back on virgin resource extraction would minimize the largest category of waste.[69]

Even though recycling offers advantages, grassroots advocates, environmentalists, and social justice activists would be remiss to stop with recycling: deeper structural changes are required to truly address the rubbish crisis. Recycling is the last line of defense in the environmental movement's struggle against destructive and needless garbage; if it gets halted or blocked, municipalities have no choice but revert to burning and burying all wastes. Cutting back on production and consumption, and reusing materials—in addition to recycling—would lead to a much more significant reduction in waste and present a much longer-term solution.

Compactor trucks waiting to dump at an incinerator. (Courtesy of Getty Images)

8

The Corporatization of Garbage

"Where you gonna go? Who you gonna get? No one's gonna service you!" yelled a Barretti Carting Corporation representative into the phone before slamming the receiver down. This was customer service in New York City under the notorious Mafia-led garbage cartel. Clients who didn't like their rates or wanted to switch haulers had a lot of trouble on their hands. Fussy patrons might suffer verbal abuse, or they might get a visit from their garbage collector, which wasn't always a pleasant experience. "Hey, why send a poodle when you can send a pit bull?" a Barretti enforcer explained.[1]

Since the mid-1950s, when local officials handed over commercial waste collection to private haulers, New York City's garbage industry had been dominated by a Mafia-led cartel (the city government's Department of Sanitation has continued to collect residential garbage). Under this regime, most shops, restaurants, offices and large apartment buildings suffered the fate that Barretti's customers were subjected to. The Mob was in charge of garbage and, it seemed, no one could stop them. Some of the Big Apple's trash carting businesses were more directly Mafia-connected, like Barretti, but many were not; yet if they wanted to run their collection trucks down New York City streets, they had to join the cartel. The payoff for being a member was guaranteed customers and a healthy income—in other words, protection against the market forces that might drive prices down and companies out of business. Estimates put the amount that the cartel overcharged its customers in the hundreds of millions of dollars each year.

All this ended rather abruptly in the mid-1990s when the Mafia's New York City garbage monopoly was destroyed. The confluence of forces behind the cartel's demise included a police crackdown on Mob activity and, perhaps more significantly, a major restructuring in the garbage industry. No longer was refuse treatment simply a service municipalities conducted themselves or contracted out to quaint mom-and-pop hauling companies. Multinational trash corporations, started in the 1970s and 1980s by a handful of innovative refuse firms, were seizing control of the garbage trade.

Two early players in the corporatization of garbage were Waste Management Inc. and Browning-Ferris Industries (BFI). They spearheaded the consolidation of the industry, buying out smaller firms in towns across the Sunbelt, then spreading north and into international markets.[2] In building their empires, WMI and BFI brought the dark and fetid world of rejectamenta into the realm of billion-dollar revenues and the New York Stock Exchange, and in the process radically remade garbage handling.[3] Among the most significant changes the rubbish conglomerates wrought were the domination of the waste market by a few large firms, the reinvigoration of the sanitary landfill and the exporting of garbage. All of these changes happened in rapid succession in New York City during the 1990s as the Mafia was driven out of the garbage trade.

With the corporations leading the way, landfilling saw a major revival in the 1990s, following a decade of decline. After all, burial was the most profitable form of waste treatment, so the corporations couldn't let it go. With their massive budgets, they could afford to purchase and upgrade old landfills that had been shuttered or were destined for closure under Subtitle D, the law that mandated safety standards for landfills. In 1993, the EPA passed federal regulations further tightening Subtitle D; the new rules required all future landfills to protect against groundwater and air pollution by installing liners, and leachate and gas collection and monitoring systems. These new protections, ironically, priced most public facilities and small local firms out of the market, a dynamic that fit neatly with the consolidation the giant garbage firms were pursuing.[4]

The conglomerates started buying up defunct dumps to upgrade and opening new regional disposal sites in states like Pennsylvania, Virginia, Ohio and Michigan. Here the trash corporations built colossal new "mega-fills," high-tech landfills outfitted with all the required monitoring equipment. And while the overall number of landfills continued to fall, as it had for almost a decade, actual trash burial capacity soared. Whereas yesterday's landfills could take in anywhere from dozens to several hundred tons of refuse daily, the new corporate mega-fills could handle thousands of tons per day.

To run these mammoth disposal sites economically, the giant firms had to make sure they took in the full amount of trash their permits allowed. If they received less, they couldn't maximize their profit potential. So the trash nationals had to start exporting garbage to new locations. And as the conglomerates moved into the Rustbelt they tapped areas desperate for economic development, finding local governments that were willing to accept refuse from other communities in exchange for a cut of the dumping fees. The garbage giants therefore began shuttling wastes from the city to the country, filling in unused land, abandoned mines and defunct industrial sites with the ever-abundant supply of trashed commodities.

Today, the export of wastes also takes place on a global scale. Castoffs from the United States, the world's top producer of trash, are pouring into the Global South at an astounding pace. Exporting wastes is the result of cheap labor abroad and stricter environmental controls at home. If it's prohibited or too costly to discard in the United States, companies can just send refuse overseas, and at a competitive price. Increasingly, in the post-industrial economy, goods are not only manufactured in developing countries but discarded there as well. The environmental fallout of refuse treatment is becoming ever-more invisible to Western consumers, distanced as they are from the real, on-the-ground ramifications of so much waste.

Waste, Consolidated

The trash nationals started out as modest, privately owned hauling companies that gradually expanded by purchasing other small concerns. But WMI and BFI took their trash public, offering stocks to raise revenues, diverging sharply from standard industry practice. And within a few short decades a handful of companies like Allied

Waste Industries and USA Waste Services, Inc., followed WMI and BFI into the staggeringly profitable world of mega-trash.

Throughout the 1970s and 1980s industry leaders WMI and BFI employed a strategy new to the trash trade, sometimes called the "hub and spoke" model. This vertically integrated approach generally took shape as follows: acquire existing infrastructure like transfer stations and landfills, buy or open small carting operations, edge out the competition by using predatory pricing or "low-balling" (charging less than fair market rates), then acquire the subsequently devalued local firms and dominate the market. The spokes were the collection routes and the hubs were the transfer stations, landfills and incinerators. Proprietorship of the latter meant controlling garbage flows by dictating disposal prices.[5] Owning landfills in particular made real economic sense since they generated more before-tax income than all other waste company operations combined.[6]

Former number two waste handling company BFI fired up its first collection truck in 1969 in Houston, Texas, allegedly after founder Tom Fatjo grew disgruntled with his local garbage service.[7] BFI was the first refuse company to expand across North America. Between 1969 and 1970 the firm overtook twenty separate U.S. markets and, by 1980, it had operations and subsidiary companies in most U.S. states, Canada, Australia, Puerto Rico, Western Europe and the Middle East.[8] In the late 1990s, BFI was bought by another firm, Allied, which is now the country's second largest solid waste management company, with revenues of $5.5 billion annually.[9]

Currently the largest trash corporation, Waste Management Inc. ate up the competition in markets across the United States throughout the 1980s and 1990s. Started in 1968 by Dean Buntrock in Chicago, WMI quickly grew from a local hauling company owned by Buntrock's in-laws, the Huizenga family. In 1970 WMI joined forces with cousin H. Wayne Huizenga, who was running his own trash company in Florida. June 1971 saw WMI join the roster at the New York Stock Exchange, one of the first garbage companies to go public.[10] Scaled back in recent years, at its peak in the mid-1990s WMI had operations in Saudi Arabia, Hong Kong, Thailand, Mexico and Italy. Today WMI has revenues topping $11 billion, with garbage hauling services, sanitary landfills, incinerators, hazardous materials landfills, and even a low-level nuclear dump in the United States.[11]

The story of the giants is a meandering tale of mergers and hostile buyouts that got particularly vicious in New York City in the mid-1990s. Spiked with Mafia brutality, these events were perhaps more dramatic than in other towns across the country, but nevertheless reveal the giants' formula for market takeover. Today the nationals dominate—the top three rubbish corporations now control almost 40 percent of the $44 billion industry.[12] New York City thus provides an excellent case study of how the trash corporations work.

Wiseguys and Garbage

The recession of the early 1990s meant that garbage volumes were down and prospects for future market expansion were on the wane, pushing the garbage giants to consider the heretofore unthinkable: cracking Gotham. In the previous quarter-century the trash nationals had already consolidated most U.S. markets, leaving scant room to expand except in a few regions like Mobbed-up, trash-rich New

York.[13] The city produced fully 5 percent of the country's commercial refuse and its rates were the costliest nationwide—a tantalizing $1 billion to $1.5 billion annual market.[14] The powerful conglomerates couldn't let the crown jewel of garbage slip away, so they decided to take on the Mob. Embedded in this noir narrative is the story of small businesses struggling to protect their stake in an economy increasingly dominated by the garbage corporations, themselves reliant on an official protection racket of sweetheart government contracts and police power.

Until this time, the Italian-American Mafia had extensive control over the trash trade, particularly in New York, New Jersey and Pennsylvania.[15] New York City's crooked rubbish handling stemmed not from the Italian quarters of Gotham, but instead from the city's tough Jews, who, although they did not deal in trash hauling, provided the Mafia with an innovative and successful business model. Early-twentieth-century Jewish mobsters worked in the Big Apple's trucking and restaurant industries, where they infiltrated labor unions and formed corrupt professional associations to link workers with management and unite competing firms in price-fixing alliances, all for the purpose of leveraging ever-higher fees from customers. Since most Jewish gangsters did not allow their children into the business, their Italian colleagues took over. With such an effective and profitable model in place, the rising Mafiosi only had to maintain and expand the enterprise; in the early 1950s they applied the formula to New York City's trash trade.[16]

According to New York Police Department undercover investigator Rick Cowan and journalist Douglas Century, writing in *Takedown*, the Mob understood the importance of trash in the post–World War II urban economy: "Control the flow of garbage, and just as surely as if you owned the supply of fresh water or electricity, you had an entire sprawling metropolis by the jugular."[17] The Mafia was on to something big.

In Gotham what emerged in the next half-century was a single citywide cartel based on "property rights," an intricate system of turf protection that encompassed rules restricting competition and ensuring massive profits to those who joined. Any "independent" rubbish collector who tried to take a cartel member's "stop" (customer) was urged in persuasive Mafia fashion to give it back.[18] Verbal warnings unheeded were followed by firebombed warehouses, smashed or stolen collection trucks, brutal beatings of owners and workers, and, in a few extreme cases, execution-style murders.[19]

Without fear of being underbid, cartel haulers were free to charge the legal maximum. Corrupt labor unions also helped by staging strategically timed strikes and pickets to drive up fees. This led to the highest prices nationwide: New York City averaged $14.70 per cubic yard, compared to $5 in Chicago and $4.25 in Philadelphia.[20] Come the 1990s, the estimated average overcharge across New York City was estimated as high as 100 percent, or about $500 million annually.[21] Customers were also menaced with blunt Mafia force if they tried to switch carters or resisted paying inflated fees. As a result, such prominent clients as Citibank, J.P. Morgan and HBO were held hostage, according to Cowan and Century: "high-placed executives of major corporations . . . big shots with Ivy League degrees and Brooks Brothers suits, Fortune 500 executives . . . were being gouged for their garbage rates. And they had no choice but to take this abuse from pinky-ring wearing, cigar-chomping garbage gangsters."[22] (HBO later exacted its revenge by creating the popular television series *The Sopranos*, about a Mafia family in the garbage cartel.) The illegal property rights regime, therefore, gave Mob-connected garbage companies permanent "ownership" over their customers and locations, keeping out new competitors and assuring New York City carters the highest profits in the country.[23]

As a result, the New York City Mafia's cartel held off much of the consolidation that had transformed the garbage industry in other parts of the country. By the 1980s, when small haulers in Sunbelt cities were being priced out then gobbled up by emerging national corporations like WMI and BFI, New York City still had almost 500 small local companies collecting its commercial discards alone. In the early 1990s, one area of Manhattan between 36th and 48th Streets, and Fifth Avenue and the Hudson River had ninety-five separate haulers, eighty of them with ten or fewer stops.[24]

The unpredictable patchwork system resulting from the cartel was evidence of a strangely stable, almost premodern method of self-regulation in which turf and contracts were allocated according to familial connection, Byzantine networks of shifting alliances, and old-fashioned intimidation and terror.

Mob vs. Mob

Leading the so-called legitimate industry's charge into New York City's trash market was the pioneering BFI, at the time pulling in $3.2 billion annually. In 1993, just weeks after BFI's first New York

City contract bids went out, the Mafia delivered an ominous message to the Westchester home of one company executive. David A. Kirschtel opened his front door to find the severed head of a German shepherd with a note in its mouth that read "Welcome to New York."[25] In the coming months, according to BFI, employees were harassed and followed, their trucks vandalized and stolen, "senior officers received threatening calls and anonymous mail," and two-thirds of their new customers "backed out after visits from their then-current haulers."[26] To defend *their* turf, BFI hired Kroll Inc. (a "risk consulting company") to scout out routes, and retained the services of armed guards to escort collection trucks.[27]

When BFI held a press conference announcing its intentions to be the first of the nationals to break into the New York garbage trade, the event was staged at City Hall. Flanked by Mayor David Dinkins and with the blessing of consumer affairs commissioner Mark Green, a BFI spokesman spelled it out for the wiseguys: "We're not being run out of town."[28] What gave the previously timid giant its aggressive new stance against the Mob? On top of being perceived as a clean business and garnering the tacit support of local elected officials and law enforcement, BFI benefited from a powerful connection through its then-chairman, William D. Ruckelshaus. Before serving as the first director of the Environmental Protection Agency and again as its head in subsequent years, Ruckelshaus, in his younger days, had worked in the Justice Department with none other than New York County's then-district attorney, Robert Morgenthau. This relationship provided BFI with access to a high-stakes undercover New York Police Department investigation against the garbage cartel, code-named "Operation Wasteland." In his position as DA, Morgenthau had authority over the three-year probe and was privy to its findings. By sending an agent from the DA's office to work undercover at BFI's newly opened New York branch, Morgenthau formally linked the corporation with the investigation, which assured Ruckelshaus continuous inside information on the cartel and excellent PR when the Mafia was later brought down.[29]

To court the hearts and minds of New York's garbage customers—many of whom were fed up with Mafia-style trash collection—BFI deployed a massive public relations offensive against the local haulers. Since the public hated being bullied and ripped off by Mafia-connected trash collectors, BFI had a sympathetic audience. But the cartel retaliated with its own savvy PR campaign. With the help of Edelman Public Relations Worldwide, they produced glossy pamphlets and slick television and newspaper ads attacking BFI. One of the cartel's commercials featured the image of a briefcase bulging with cash accompanied by a voice-over enumerating (ironically enough) BFI's own less-than-legal activity. According to the ad, the Texas-based corporation's record included "accepting bribes, antitrust convictions, toxic waste violations, and heavy fines for price-fixing in six states."[30] Although the locals were intent on excising BFI for their own benefit, they were right to point out that the industry leader was not so clean after all.

Despite their efforts to resist the nationals, by the summer of 1995 New York City's garbage cartel was officially wiped out. Thanks to the extensive undercover investigation by the NYPD, a whopping 114-count indictment was handed down. All the principal members of the cartel were subsequently arrested and every company and individual who was charged either pleaded or was found guilty.[31]

Even though the majority of New York City's haulers were not indicted, and a portion of them were not connected to the Mob, Morgenthau's office treated all locals as fair game. According to a deputy DA, "No one gets a clean bill of health."[32] Cinching local government's oversight of the trash industry, Mayor Rudolph Giuliani enacted a 1996 law creating the Trade Waste Commission. It gave city officials sweeping powers to reject handling licenses, whether a firm had been criminally charged or not.[33]

The trade publication *Waste News* conveyed the feeling among vulnerable haulers: "Many assume some carters who weren't indicted will be denied licenses and effectively thrown out of business."[34] This approach massively transformed Greater New York's garbage business by shutting out many locals and easing the entrance of the country's largest and most powerful trash corporations.

The End of Competition, Again

After the fall of the New York cartel, WMI and USA Waste joined BFI in Gotham for the feeding frenzy.[35] All three firms' strategies for entering the Big Apple's market were the same as they had been in other towns—the hub-and-spoke model. This system, not surprisingly, could lead to business practices of questionable legality. The giants' records foreshadowed what was to come in New York.

USA Waste, WMI and BFI all had long rap sheets, some dating back to the 1970s, with antitrust violations and tens of millions of dollars in fines.[36] WMI alone handed over more than $100 million to settle more than 200 criminal, antitrust and environmental cases in at least twenty-three U.S. states between 1970 and 1991.[37] By the early 1990s BFI executives had already been charged and disciplined for bid rigging, price-fixing, and bribing public officials in Ohio, New Jersey, Georgia and Vermont.[38] During the Vermont hearing, one worker testified that his superior instructed him to first "cut the price in half" and after the competition folded "then you double the prices."[39] Predatory pricing was to the giants what property rights were to the Mafia.

However, unlike their underworld competitors, these billion-dollar corporations barely felt the sting of such penalties and, in an asymmetrical application of the law, typically did not lose operating licenses for unfair business practices. As a result, the nationals were largely unfettered in their takeover of post-cartel New York.

The opening years of "free competition" in the mid-1990s brought a drastic fall in commercial refuse hauling prices as average fees sunk by 30 percent in one year.[40] And during this time no single carting company had more than 5 percent of New York City's commercial customers.[41] For many, this confirmed the claims of government, media, and the waste handling giants of successfully restoring competition to the Big Apple's waste trade.

This period of increased competition, however, was but a brief spell. In reality, the nationals were implementing predatory pricing to strangle the locals by driving rates below market value.[42] After surviving two decades of ruthless mergers, acquisitions, downsizing and consolidation in the trash industry, New York City's small carters were finally on the ropes. During this time, single contracts could go for a fraction of their previous price, like at the World Trade Center, where a new deal was struck for 75 percent below its former rate.[43] Kalman Gregor, owner of Gregor Carting Corporation, lost his customer of forty years, the Manhattan department store Lord and Taylor, after BFI underbid him by 25 percent. Gregor's three-truck, four-man operation was devastated by the lost revenue: "It carried us. It was the life stay of the business."[44]

With many of New York City's small haulers stunned and struggling, the nationals moved in for the kill. The late 1990s became a period of intense consolidation as WMI seized the lion's share of the market, with BFI not far behind.[45] Some predicted that New York was headed back from whence it came, according to an article in the journal *Social Justice*: "Thus, the criminal cartel monopoly was replaced with a two-firm oligopoly."[46] Richard Schrader, who supported the entrance of the nationals when he served as a consumer affairs commissioner under Mayor Dinkins, spoke out in 1998 about the mounting lack of competition in the industry: "The business community felt the old mob cartel was unfair and charged too much. But now there's no one to play off against the new monopoly. As prices go up, business consumers will be looking for competition, and the fear is that it won't be there."[47] Schrader was right. By the end of the 1990s the number of carters licensed to operate in New York City fell sharply, almost 70 percent. And by 2002, the city's disposal rates were up 40 percent from six years earlier, rivaling the inflated fees of the Mafia's cartel.[48]

After taking over the Big Apple's trash hauling market, the national corporations went on to remake refuse disposal practices in the U.S. Northeast. In the rural areas close to New York City and other urban centers nationwide, the giants have pioneered the controversial practice of exporting garbage.

Location Is Everything

In 2001, New York City closed its last large-scale disposal site, the hotly contested Fresh Kills Landfill on Staten Island. The capping of the dump was made possible by the garbage nationals' mid-1990s takeover of the city's rubbish hauling market. In addition to buying up smaller companies, as they entered the market the garbage corporations used their massive capital to acquire and build key regional transfer stations, incinerators and landfills. By 1997, WMI, USA Waste and BFI already owned more than enough nearby disposal capacity to handle all of the Big Apple's castoffs.[49] This corporate trash infrastructure provided the city with an alternative, allowing Mayor Giuliani to permanently shutter Fresh Kills. (The site was briefly reopened to dispose of the debris from the 9/11 World Trade Center attack.)

Significantly, much of this new network for managing the Empire City's refuse was located in its immigrant, working-class urban neighborhoods and cash-strapped rural regions across state lines. Some reasons for the changing geography of rubbish can be traced to the garbage industry's awareness of the power of public resistance. A key industry study from 1984 explained: "All socioeconomic groupings tend to resent the nearby siting of major facilities, but middle and upper socioeconomic status possess better resources to effectuate their opposition." The perfect locations for disposal operations, according to the report, were low-income rural areas with populations of less than 25,000 that were older and had high school or less education.[50]

Thus, facilitated by the garbage corporations' consolidation of New York City's trash market, Staten Island's majority white and Republican residents—who would go on to support Giuliani in a second term—shed their putrefying, filthy burden onto less politically powerful rural areas. They also shoved the mess onto neighborhoods like New York City's South Bronx, home to a sprawling new transfer station, a vital component of the new trash-exporting framework.

The South Bronx's Harlem River Yard Transfer Station is a mammoth fourteen-acre, $25 million facility, opened by WMI after the firm received the first city contract to export Gotham's household garbage in 1999.[51] When operations at the Harlem River Yard commenced, the predominantly African American and Latino low-income neighborhood was already saturated with thirty-five waste transfer stations and had some of the highest asthma rates in the country.[52] Community groups like Sustainable South Bronx and *Nos Quedamos* (We Stay) opposed the project, as did the South Bronx Clean Air Coalition (SBCAC) who filed a lawsuit to stop it. According to the SBCAC, what WMI and obedient local officials were committing was environmental racism. "It's like the wild, wild West and the waste companies have the run of the streets," SBCAC's Carlos Padilla told the *New York Times*.[53]

Despite this opposition, WMI has thus far prevailed due in part to officials like Martha Hirst, a city deputy sanitation commissioner. In

2000, she defended the siting of facilities in neighborhoods like the South Bronx: "The transfer stations are to some degree clustered in those areas and that's appropriate, that's where the city map, the land-use regulations say they ought to be."[54] WMI also found an ally in former state attorney general Dennis C. Vacco who, after leaving his public post in the late 1990s, became regional vice president for the firm.[55] The Harlem River Yard handles thousands of tons of refuse per day, and along with the neighborhood's other garbage processing facilities, as much as 40 percent of New York City's privately collected refuse now flows through the South Bronx on its way out of town.[56]

A hefty surge in truck traffic has been another outcome of New York City's increased reliance on the garbage giants to handle its wastes. Until 2000, most of the city's rubbish was processed at eight marine transfer stations, with barges as the main mode of transportation. But the nationals do things differently. While they use some rail, companies like WMI rely mostly on trucks. Annually, there are over 250,000 trips made by exhaust-belching diesel rigs hauling refuse through city streets, plus an additional 250,000 long-distance truck trips to deliver garbage out of state.[57] Locally, this contributes to the more than twenty asthma deaths each year in the South Bronx alone.[58] More broadly, all those extra vehicles on the road are spectres of a grim future: NASA scientists recently concluded that soot, predominantly from diesel engines, is causing as much as a quarter of all global warming.[59]

Once New York's trash crosses state lines it usually gets deposited in one of the region's mega-fills. Most of these new, improved dumps are owned by the garbage giants and located in rural areas accessible from dense urban centers. New York City's wastes get shipped mostly to Pennsylvania, Ohio and Virginia. These states have become trash hubs because of their geographical proximity to mid-Atlantic cities, large swaths of undeveloped land and local economies hobbled by deindustrialization. Economically hollowed-out Rustbelt towns are often lured into approving disposal sites in exchange for a percentage of fees.[60] Significant, too, are these towns' demographics.

It is no coincidence that, as with other trash importing states like Indiana, Michigan and Illinois, in Virginia's counties with mega-fills, "residents are on average much poorer, less well educated and more likely to be African American than the average Virginian." Hence, in the eyes of the waste industry, such facilities are located at points of least resistance.[61] These communities are increasingly organizing against the environmental injustice thrust upon them, disproving standard industry assumptions about their passivity. Nevertheless, they continue to be recipients of a disproportionate amount of discards and they continue to be targeted by the waste industry.

Consume Locally, Dump Globally

Still vivid in many Americans' imaginations is the *Khian Sea*, the infamous garbage barge that, in 1986, set out on a twenty-seven-month voyage down the East Coast then through the Caribbean, illegally dumping a third of its cargo of toxic incinerator ash on Haiti's beaches before slipping away under cover of night. Lurking persistently on the dark seas, unwanted yet irrepressible, the ship and its cargo periodically emerged in news accounts of its futile search for a disposal site. Then, after getting turned back from ports in Africa and South Asia, the vessel mysteriously offloaded its remaining 10,000 tons of ash somewhere between the Suez Canal and Singapore.[62]

The saga of the *Khian Sea* triggered anxiety in the United States about the sustainability of its garbage disposal methods; the load represented merely one month's worth of incinerator slag from a single city. But the event also illuminated a larger, growing pattern of dumping waste in impoverished, less politically influential countries. Larry Summers, while chief economist at the World Bank, argued that "the economic logic behind dumping a load of toxic waste in the lowest wage country is impeccable, and we should face up to it. . . . Underpopulated countries in Africa are vastly underpolluted . . . ."[63] It sounds appalling to the rest of us, but his observation was already common knowledge in the trash trade.

The same market incentives and political perceptions that drive domestic rubbish exporting also fuel the growing trend of sending wastes overseas. Some shipping companies that bring consumer goods into the United States have taken up rubbish handling. Instead of returning with empty vessels, they fill their cargo containers with U.S. wastes, which they then sell to recycling and disposal operations in their home countries.[64]

Perhaps the most malignant and abundant exported trash today is "e-waste," obsolete electronics like VCRs, CD players, televisions, computers, cell phones and fax machines. Rich in heavy metals like lead, cadmium, mercury and zinc; toxic solvents; PCBs (polychlorinated biphenyls); and other hazardous materials, e-waste is virulent stuff. And there's so much of it. The electronics industry, most notably producers of cell phones and personal computers, has brought built-in obsolescence to dizzying new heights. Not only is it practically impossible to get replacement parts for a cell phone, anyone caught talking into a bulky model that's more than a dozen months old risks public humiliation. And forget finding technical assistance for a computer that's been around for more than a few years. The pressure and enticement to consume more electronics at ever-faster rates, thanks to unrelenting fashion and technological innovations, are fueling a vast new garbage boom.

The current industry-projected lifespan for personal computers is about two years. Today, there are over 100 million people using PCs in the United States alone. Since 1997 the country has produced a tidal wave of junked laptops, monitors and hard drives estimated in excess of 300 million. And, in addition to PCs, 3.2 million tons of other high-tech hardware gets tossed every year.[65] Most defunct monitors contain between four and eight pounds of lead and are now the source of 40 percent of all lead in U.S. landfills.[66] Some 70 percent of all heavy metals found in U.S. landfills come from discarded electronic components like wires, circuit boards and metal casings.[67]

Many foreign shipping firms now haul a significant portion of U.S. e-waste to the Global South for processing and disposal. And with cheap labor so abundant in these impoverished regions, sending e-junk out of the country is economical. One EPA study found that recycling a computer monitor in California cost ten times more than shipping the rubbish to China for recycling.[68] Moreover, the United

States has yet to ratify the Basel Convention, an international accord that regulates the trafficking of hazardous wastes from industrialized to developing countries.[69] This combination of more closely monitored dumping at home, low-cost labor abroad, and lack of regulations on the wastes streaming out of U.S. ports has led to a thriving rubbish trade in countries like India, China, South Africa and the Philippines.

Some reports reveal that between 50 and 80 percent of e-wastes collected for recycling in the United States are sold to dealers who export it to developing countries for dismantling.[70] Even if American consumers pay a premium to recycle their PCs, there's no guarantee of safe, proper treatment.[71] Disposal and recycling of e-waste in developing countries is a sometimes illegal, typically unrestricted, environmentally nefarious endeavor forced on the poorest workers who, according to a *Seattle Times* report, use "hammers, chisels and bare hands to break and sort computer parts." While the idea of recycling discarded electronics might sound good, the reality in many overseas plants is a nasty coalescing of human health hazards and environmental destruction.[72]

In the town of Guiyu in China's Guangdong Province, migrant laborers deconstruct and melt down mountains of computer parts, the remains of which are often dumped into nearby rice fields, irrigation canals and along waterways. The groundwater in the area has become so polluted that potable water must be hauled in from thirty kilometers away. Elsewhere in the country, dumped cell phones are leaching brominated flame retardants from their plastic components, poisoning groundwater and soil.[73]

In a computer recycling facility in New Delhi, laborers salvage copper and silver from circuit boards by dipping them into barrels full of acid. These workers wear almost no protective gear, only thin rubber gloves if they're lucky. Once the process is complete, the company dumps the exhausted chemicals directly into the local sewage system and burns the plastic boards in the open air. The plant's owner explains that he gets much of his debris from North America, even though the importation of e-waste is illegal in India. Because of the industry's underground nature, statistics on how many spent electronics enter the country are hard to come by. Posing as domestic scrap buyers, a local watchdog group called Toxics Link New Delhi investigated three Indian port cities. In just one location, Ahmedabad, in the northwestern state of Gujarat, the group found that thirty tons of computer rubbish arrived monthly. Without adequate funding and the political will to keep tabs on the waste trade, importers and processors will continue to operate and expand.[74]

More familiar categories of wastes are also shipped overseas for recycling and disposal. In 2002 the United States exported to China 2.3 million metric tons of scrapped iron and steel and the equivalent of about 14.5 billion used plastic soda bottles. All those castoffs are fueling a roaring overseas recyclables market: between 1997 and 2002 the value of U.S. waste exports to China surged from $194 million to $1.2 billion. One scrap dealer, America Chung Nam, sent out more containers from U.S. ports in 2002 than did DuPont, General Electric, and Phillip Morris (now called Altria) combined. It turns out that garbage is one of the great cultural exports of the United States.[75]

Ever-more monolithic rubbish graveyards and the egregious exporting of trash are the result of more refuse, but also the outcome of inadequate regulatory laws. By not taking the crucial step of restricting the production of toxic substances and needless waste, governments have fostered a situation in which manufacturers continue to make and market a boundless supply of rapidly obsolete commodities.[76] Without limiting the manufacturing of so many wastes the piles of wrappers, cans, plastic and outmoded electronics will inevitably mount. And if these discards can't be disposed of here, in our backyard, they will simply get shipped overseas.

Until we address the root causes of America's massive and malignant wastes and work to reduce trash before it gets made, resolving the garbage crisis will remain elusive. Describing the problem of housing in the nineteenth century, Friedrich Engels wrote, "The bourgeoisie has only one method of settling the housing question. . . . The breeding places of disease, the infamous holes and cellars in which the capitalist mode of production confines our workers night after night are not abolished; they are merely *shifted elsewhere*."[77] Official solutions for waste today are the same. But no matter how much it is processed, shipped, dumped, burned or buried, garbage never really goes away.

Trash taken from Dumpsters behind fast-food restaurants. (Peter Garfield, 2004)

9

Green by Any Means

> "I find that everything I see is garbage. . . . I went to a new restaurant last week, nice new place, you know, and I find myself looking at scraps of food on people's plates. Leftovers. I see butts in ashtrays. And when we get outside."
>
> "You see it everywhere because it is everywhere."
>
> "But I didn't see it before."
>
> "You're enlightened now. Be grateful," I said.
>
> DON DELILLO, *Underworld*

Just up the Batcave's inclined concrete driveway lies a compact backyard. The rear garden of this dwelling in Oakland, California, is no longer a patch of fertilized, clipped sod, but has been transformed into a thriving vegetable garden with squash, potatoes, strawberry plants, lettuce, flowers, beehives and a handmade solar oven. Surrounded by neighboring single-family bungalow style homes, the Batcave, as its denizens refer to it, is a low-waste abode created and occupied by a crew of political and environmental activists. The Batcave's garden is not only where the group grows a sizable portion of its food, but in this quiet oasis they process much of their waste.

A retooled plumbing system channels all the house's "gray-water"—wastewater that flows down sinks and shower drains—into a homemade wetland system. A series of cascading pools filters out impurities as the wastewater passes through rocks, sand, cattails and water hyacinths, rendering a prolific supply for garden irrigation. A few short steps from the back door sits a series of organic waste composting bins. One has active red worms that process carrot tops, onion skins and moldy bread into a rich soil amendment in a matter of days.

"We only throw out one small bag of garbage for the whole house of eight adults each week," explains Tim Krupnik, Batcave resident. Krupnik and his housemates practice a "strict conservation ethic," growing much of their own food, dumpster diving and buying almost nothing new to maximize their use of resources. The house even boasts a compost toilet, where human excrement can be collected for processing into fertilizer.[1]

While this may not be enticing or realistic for the average eco-friendly American, the Batcave proves that a low-waste lifestyle is possible. Small-scale, individual rubbish reduction and reuse efforts like the Batcave's are important visionary projects that demonstrate the viability of reconfiguring production, commodity circulation, consumption and wasting. By now it is commonly recognized among mainstream and radical environmental groups alike that discards should not be treated *after* they get made, but instead society should generate less waste. Advocates for addressing the root causes of rubbish vary widely in their approaches and implementation of alternatives.

A plethora of waste solutions are available, other than those offered by industry today. In addition to changes made based on individual choice, the concept of "green capitalism" aims to keep mass consumption chugging along while business and industry take up more costly environmentally sound measures on their own, without government involvement. An effort at reducing discards using state intervention to influence changes in production is called "extended producer responsibility." In this system, manufacturers are held financially accountable for the consumer wastes their products generate. Some activists and policy analysts advocate a return to refillable beverage containers as a low-cost way of reducing a huge category of discards. The goal of "zero waste," another alternative, is to create mandated garbage reduction quotas, indirectly limiting production with the aim of eliminating waste altogether. All of these approaches stem from a common recognition that something is dangerously off-kilter, but these solutions differ in crucial ways that warrant closer scrutiny.

Refuseniks

People unwilling to wait for official solutions have always made meaningful contributions to imagining different ways of living. The following admirable yet lilliputian efforts help point the way forward in the realm of garbage.

Working as a free agent, San Francisco resident Caycee Cullen encourages people to stop consuming all those to-go cups and reuse empty food jars to drink from instead. (So many cafes and restaurants have switched to disposable dishware that, even if one doesn't order takeout, reusable ceramic cups and plates are increasingly scarce.) Americans consume more than 125 million disposable cups every day.[2] Cullen distributes clean, empty jars along with an illustrated instruction 'zine that notes the importance of keeping the container's original lid, to prevent leaking. The last page features a drawing of discarded paper cups piled high juxtaposed to the lone glass vessel with the lines: "The Jar keeps my daily contribution to the landfill low. One jar, not this waste."[3]

On a larger scale, the grassroots practice of "freecycling" is gaining momentum in towns across the globe. Started in 2003 by Tucson resident Deron Beal, freecycling matches people trying to purge with those in need of objects ranging from telephone poles to personal computers. These unions take place on local branches of the main freecycling website with the primary stipulation that everything must be given for free, the goal being to keep materials out of the landfill.[4] More immediate than classified ads, the Internet facilitates the quick linking of people and goods for reuse, eliminating time and communication delays that often lead to wasting.

Old-fashioned salvaging can also put a serious dent in trash production, as the owners of Berkeley's Urban Ore have learned. Thanks to an unusual municipal contract, Urban Ore has a license to glean discards from the city's dump. The for-profit business sells its haul at a depot it calls "Eco Park," a former factory surrounded by a giant outdoor lot. Countless rows of old white, pink and yellow toilets, sinks and bathtubs line the yard's western quadrant. Nearby, delicately paneled doors ripped from remodeled Victorians stand vertical and neat in a series of aisles. Near the back of the massive grey building are several clusters of upright windows leaning against each other, translucent and glinting in the sun. Inside, a wall of file cabinets butt up against a mound of precariously stacked chairs and everything from building materials, tools and old light fixtures to spatulas, desks and electronics are for sale.

Established in 1980, Urban Ore is dedicated to rerouting discarded goods away from the landfill and back into circulation. Mary Lou Van Deventer, co-owner of the enterprise, explains: "When we salvage for reuse and we save something in its already manufactured form, we conserve the material, the cultural value, and the energy that went into producing that object. Now, if the same object is shredded up and bashed and remanufactured into something new, it's good for it to be recycled, but it isn't nearly as good as if that value had been conserved in the first place by using the object as is."[5] Urban Ore is the kind of operation that could be duplicated everywhere. Municipalities could be required by law to contract with salvaging companies, creating an improved version of common nineteenth-century scavenging and reuse practices.

These attempts at reducing and reusing, and the countless others not mentioned here, reveal the wide range of grassroots alternatives to burning, burying and official forms of recycling.

Having It Both Ways

Green capitalism is a waste-cutting solution currently gaining popularity among environmentalists, policy wonks, enlightened entrepreneurs, yuppies and New Agers alike. A product of the early 1990s and a kindred spirit to "socially responsible" business, green capitalism forges environmentally sound approaches to production that keep business and industry from poisoning and killing the planet. Central to this "new industrial revolution," as its proponents call it, is no government intervention and a belief that with the right tweaking, high consumption levels can continue unabated. While green capitalism's goal of taking care of nature is commendable, and while it rightly asserts that production processes need to change, the means by which it aims to achieve these goals must be interrogated to ascertain their viability.

Some of green capitalism's most vigorous promoters are the upscale gardening chain-store owner Paul Hawken and the eco-minded architect William McDonough. Both have published books outlining their ideas and strategies for creating greener business, at the center of which lies design. Capitalism is not the root of environmental decline, green capitalists proclaim. The real problem stems from poor product and manufacturing design, which inevitably leads to huge amounts of wasted natural resources. Shunning government intervention, green capitalism's goal is for individual companies to voluntarily back-engineer manufacturing so that wastes are designed out of the process; castoffs would either break down harmlessly into the natural environment or be recycled infinitely in a nontoxic manner. It's a seductively simple idea. But there's a catch: how will this industrial transformation take place with no reduction in consumption and on a purely voluntary basis?

Continued high consumption levels are possible under green capitalism if companies redesign their products to be endlessly reincarnated or to harmlessly disintegrate (a new twist on disposability).

Under this system, markets can theoretically continue expanding, bypassing the perceived threat to business of reduced consumption that is so often associated with a pro-environmental agenda. The problem here is that ongoing mass production and its accompanying waste will persist.

This outcome is apparent with new forms of plastics. Current iterations of "bioplastics" (endorsed by McDonough and others) like those made from starch, soy and hemp may require huge amounts of water and energy to manufacture. They are also likely to promote monocropping and increased use of chemical fertilizers to ensure a uniform and reliable feedstock. This would further wipe out biodiversity and pollute water and soil. Additionally, increased demand and higher prices for crops to feed the plastics sector could impact the food supply, since those with the strongest purchasing power get the goods. Journalist George Monbiot argues: "Those who worry about the scale and intensity of today's agriculture should consider what farming will look like when it is run by the oil industry."[6]

According to environmental writer Daniel Imhoff, vegetable-based polymers are in theory better than the old resins, but "bioplastics are fundamentally a techno-fix to support our existing habits without changing the lifestyles of our convenience/consumer society."[7] And as we've seen with dumping, burning and recycling, although techno-fixes may appear to work in the short run, they are inadequate as long-term solutions because they leave intact the underlying structures that produce so much waste.

Green capitalism's other major weakness, a strident rejection of state intervention and regulatory laws, means that companies must make these changes because they believe it is the right thing to do. Furthermore, since going green often involves increasing expenses, business leaders must not only feel a sense of moral obligation, they must be willing and able to take actions that will reduce profits.

Addressing this issue in his book *Cradle to Cradle*, McDonough explains that he first establishes whether or not a potential client can become what he calls "eco-effective." "If they are firmly entrenched in . . . the grip of an ism (pure capitalism)—they might consider moving production to a place where labor and transportation are as cheap as possible, and [we] end the discussion there. If they are committed to a more stable approach, however, we press on."[8] In other words, if individual producers and businesses are able to incur greater costs, then they can go green. But if a company simply must compete in the marketplace, McDonough and his coauthor Michael Braungart cut them loose. Since capitalism is bound by competition's coercive powers—after all, getting an edge on the competition by producing commodities at ever-lower prices is the root of so much resource extraction and wasting—McDonough and Braungart drop the ball at the most crucial moment.

When individual companies assume the added costs of going green, they risk losing business to rivals not saddled with the extra expenses. For the eco-friendly manufacturer, production can easily result in more expensive commodities that can't hold their own in the marketplace.

In reality, business and industry must have unfettered access to natural resources to compete effectively. In an amoral assessment, the market's sole purpose is accumulation for accumulation's sake, meaning that its only relation to nature is accumulation.[9] To transform the relationship of business and manufacturing to nature requires the transformation of capitalism. Green capitalism's aversion to legally mandated controls on production is a reactionary tendency that may buffer its environmental agenda in the eyes of old-line industry but ultimately precludes meaningful change. As has been proven unequivocally over the last thirty years, corporate self-restraint does not work, especially when it comes to preserving the health of natural systems.

Without regulatory oversight, companies can adopt an ecologically responsible façade without making real systemic changes. Over the last several years, many companies have appeared to go green, but beneath the surface the exploitative apparatus churns away. Just because Ford hired McDonough to rebuild one of its River Rouge factories—where he famously installed grass on the roof—doesn't hinder the automaker from turning out some of the least fuel-efficient vehicles on the road, like its F-150 trucks and Expedition SUVs. Ford now exploits the "eco-effective" River Rouge plant as a tool to greenwash its image while continuing to market its gas-guzzling, carbon-dioxide-spewing hulks.

The same is true of corporate self-governance in the realm of garbage. Coca-Cola pledged a 25 percent recycling target in the early 1990s when much public attention was concentrated on the issue. As political pressure and consumer focus shifted, without any government accountability and no restrictions on the use of the recycling symbol, Coke found that it didn't need to actually recycle to maintain its green image among consumers. In this permissive climate, Coke stopped using recycled plastic in its bottles altogether in 1994 and suffered few consequences, none of them legal or lasting. In 2001 the company was faced with renewed public calls for more recycling, but feeling little pressure to aim high, it promised to use a meager 2.5 percent reprocessed resin.[10] In addition, Coca-Cola's biggest plastic PET bottle supplier, South Eastern Container, recently upgraded two factories that now output 60,000 brand-new half-liter soda bottles every hour, "which the company is claiming as a world first."[11] So much for the moral imperative.

The state has always been central not only in managing populations in times of rebellion, disease, and economic or natural disaster, but also in stimulating capitalist expansion. No major developments in U.S. industry have taken shape without direct or indirect government assistance—from nineteenth-century railroad construction and the massive reorganization and streamlining of industry during World War II to today's corporate welfare and the ongoing subsidy of waste management. Since anti-environmental free marketeers have relied so extensively on the state and public funding, why should the government not intervene to responsibly steward ecological resources? If it can muster its armies, police, massive budgets and political power on behalf of U.S. industry, why should the state not gather those forces to protect the environment?

Just as the U.S. government has acted to promote industry, it has also been forced by social movements to make structural reforms that business hates (even if these changes benefit business in the long run) yet that are good for average people—reforms such as social security, welfare and food and drug inspection. Programs such as these prove that the state can effectively enforce restructuring that serves the public good.

Waste Not . . .

In a striking move, the conservative German government of Helmut Kohl passed the landmark 1991 Packaging Ordinance, a law that shifted the burden of collecting, sorting, recycling and disposing of packaging wastes away from taxpayers and onto manufacturers. A version of "extended producer responsibility," the ordinance forced industry to administer and bankroll a separate refuse handling system just for packaging. Amid fierce resistance from business and cheers from environmentalists, the Packaging Ordinance marked a serious intervention by government in the decisions of production, disallowing industry's full externalization of commodity waste costs. Not without its limits and flaws, Germany's still-active law has served as a model program for addressing the single largest component of the discard stream—packaging.

The mandatory nationwide ordinance allowed the manufacturing sector to formulate its own methods for handling trashed wrappers and containers; the result was what came to be known as the "green dot" system. Still in operation, the program is overseen by the environmental ministry and administered by an industry-led entity called Duales System Deutschland (DSD). The procedure is simple: After consumers make their purchases, they can discard the wrapping at the store, or, once home, they can place all their spent packaging into a yellow bin or bag that gets collected at the curbside by DSD.[12]

DSD financing comes from the country's private producers, who are required to pay a licensing fee proportionate to the cost of handling their specific packaging material based on weight and type. To signal payment, these manufacturers get to stamp their cellophane, boxes, bottles and cans with the green dot symbol—two arrows swirling together—an image that not only helps consumers sort their discards, but also, like the recycling symbol in the United States, generates excellent eco-PR.

Germany's program was the first successful large-scale public implementation of extended producer responsibility (EPR). A new spin on waste reduction strategies proposed in the 1970s, EPR has gained popularity in the past decade and aims to shift from the public to manufacturers the duties of managing discarded commodities like packaging and the mushrooming piles of e-waste. The underlying idea is to persuade companies to generate less from the start. If they have to pay to handle and treat these wastes, the logic goes, producers will ultimately choose to create fewer disposables.

According to economist Frank Ackerman, the 1991 ordinance "jump-started producer responsibility—it was proof of possibility." Ackerman also explains that in the face of the German program's indisputable success, government's active role in curtailing waste production could no longer be discredited: "It was such a striking example of going from nothing to accomplishing this complex task. The government decided to do it and did it, proving that producer responsibility could actually be implemented."[13]

The green dot system invalidated opponents' predictions of economic cataclysm and claims that the state could not successfully regulate industry for environmental protection. Hailed a success, the program significantly cut packaging consumption, boosted recycling, and sustained the use of refillable bottles. In the ordinance's first half-decade, manufacturers used 7 percent less packaging; by contrast during that same period, U.S. container consumption grew by 13 percent.[14] Obligatory recycling targets under the Packaging Ordinance, which increased incrementally over time, have kept DSD on track. As a result, recycling for packaging has surged by 65 percent since 1991.[15] And the ordinance's requirement that at least 72 percent of the country's beverage containers be refillable has guaranteed reduced consumption by requiring reuse.[16] The program has proven so beneficial that, since the mid-1990s, fifteen European Union countries as well as Korea, Taiwan and Japan all enacted variations on Germany's Packaging Ordinance.[17]

Unfortunately, the German system has its downsides. In more recent years packaging production in the country has increased, despite the green dot program.[18] As of 2003, Germany was the largest single market for packaging throughout Europe.[19] So, while producers may be financing the handling of their packaging wastes, they are not deterred from producing these discards in ever-greater amounts. This reveals the system's limits; mere encouragement of rubbish reduction is not enough. While Germany's system forces manufacturers to internalize some refuse treatment fees, more accurately distributing production costs, it has not brought changes that directly cut waste.

Among its other shortcomings, the green dot signals only that a producer has paid its fees to DSD, guaranteeing neither that a package was made from recycled materials nor that it will be reprocessed. What's more, recycling is defined so broadly under the green dot system that it includes the channeling of used plastics to the steel, petrochemical and oil industries for burning as "fuel." And since recycling in Germany is just as susceptible to market fluctuations as in the United States, materials that get collected by DSD might easily end up incinerated or landfilled instead.[20]

Despite these weaknesses, the daily sifting of one's discards contributes to more widespread popular consciousness of the value of trashed materials. The German green dot is but one element in a nationwide household trash-sorting culture. In addition to segregating most packaging wastes, residents separate recyclable paper, glass and metal, while they store food castoffs in special brown bins. Consumers return empty milk and beer bottles to the supermarket and whatever is left over (not much) gets deposited into a black refuse can.[21] In the United States, such setups are not entirely unheard of. San Francisco's recently revamped municipal system calls for extensive household segregation of castoffs—sending bottles, cans and paper to recycling centers and food scraps to a compost facility, with the much reduced remainder going to the landfill.[22] San Francisco's new program is a sharp improvement on previous methods and, like the German system, it fosters an awareness of discards as useful, not just spent and dirty waste.

In their own way, these comprehensive programs address the significance of individual choice in the creation of waste. While the underlying structures that enforce garbage are paramount, individual practices are not entirely irrelevant. In the last century the majority of Americans have gleefully accepted, even embraced, a high-waste lifestyle. And in some ways it's understandable. There are real conveniences offered by disposable products like paper cups and diapers, while seductive packaging and entrancing new commodities are hard to resist for many shoppers. There's no doubt that bulk bins, canvas shopping totes and used plastic bags drying in the dish rack can seem like part of a more pinched, less spontaneous lifestyle. Consuming sexy packaging and the latest styles is, for large parts of the population, an intensely fun and gratifying experience. And, while people might feel bad about throwing things away, they may also get a type of pleasure from tossing out the old and bringing in the new.[23]

These realities, constantly reinforced in mainstream society, can present cultural and political obstacles to change. Creating an atmosphere where expanded reforms protecting human and environmental safety and health seem normal and even hip—witness programs like those in Germany and San Francisco—can foster a shift in thinking about consumption that counters the dominant line. Recycling unleashes various forms of imaginings; rather than just the misperception that it actually works as a long-term solution, recycling—crucially—also opens the political and cultural imagination to other creative possibilities.

. . . Want Not

A more sustainable solution to the flood of trash is the refillable bottle. Cast aside by major U.S. manufacturers in the 1970s and 1980s, this method of reusing packaging is profitably employed in parts of Western Europe, Latin America and Canada today, often by the same North American firms that reject using it at home. Stimulating local and regional economies, refillable bottling plants create more jobs while massively reducing packaging wastes.

The way the refillable system works might still be familiar to some: the thick bottles usually require a deposit and are returned to stores once they're empty. From there, beverage makers take them to a washing and refilling facility. After bottles are sanitized and an electronic "sniffer" has weeded out any contaminated units, the containers are filled on an assembly line at a pace equal to that in factories using brand-new packaging. Once refilled, the bottles are shipped to retailers and the cycle begins anew. Today refillables are used on average twenty times before getting tossed.[24]

Refillables have stuck around in other countries for two reasons: regulatory laws and economics. Once ubiquitous in the United States, refillables still hold the vast majority of bottle markets in Denmark, the Netherlands, Germany and Finland, and in Ontario, Quebec, and Prince Edward Island in Canada because their governments have enacted measures that require and promote reusable containers. Many of these laws were passed in the 1970s to prevent throwaways from needlessly clogging disposal sites. Beverage makers have also continued using refillables in Latin American countries like Mexico, Brazil and Argentina because these bottles cost markedly less than disposable packaging. And producing drinks that sell at a lower price opens more markets to buyers with lower incomes.[25]

Coming typically in glass or thick PET, this washable, reusable form of packaging has slashed garbage output by an estimated

380,000 tons annually in Denmark, while in Finland garbage has been reduced by an estimated 390,000 tons.[26] As of 1998, the annual per capita output of packaging wastes in Finland was half that of other European Union countries, most of which don't mandate the use of refillables.[27] The larger environmental impact of recirculating packaging includes dramatic reductions in greenhouse gas and carbon monoxide emissions, and reduced consumption of water and energy.[28] The positive effects extend into the creation of jobs; more workers are needed when companies use refillables than when they use one-way containers. If Germany switched entirely to reusable bottles, 27,000 new jobs would be created.[29] Refillables also stimulate local economies, because the system requires processing in decentralized bottling plants.

And, contrary to the dominant cultural narrative in countries like the United States over the last thirty years, people still like bringing back their empties. According to a recent Gallup poll, Finnish consumers preferred buying beer in returnable bottles by a margin of almost 80 percent, while 94 percent favored soda in reusable containers.[30] The majority of Germans—69 percent—want to buy their refreshments in returnable vessels. But it's not just consumers who like less packaging; the Quebec Brewers Association says its members' loyalty to the reusable stems from its unbeatable low cost and high customer participation. A group of regional Canadian beer makers prefer the returnable system so much that they joined in a formal agreement to continue using refillables.[31]

Despite all the upsides, the reusable is under assault from government and business in many European and Latin American countries.

The European Union has challenged Denmark's and Germany's laws as obstacles to free trade. The outcome of this offensive appears to be a gradual eroding of each country's refillable system.[32] Many drink makers are also antagonistic toward refillables, constantly angling to ditch the practice so they can centralize bottling and distribution and market the higher-priced disposables. But another culprit is supermarket chain stores like Aldi and Wal-Mart. These companies often refuse to stock refillables because of the extra space and labor that handling them requires. As these retailers take over markets formerly served by small local shops, they exercise their colossal market power over the future of refillables. If there are no laws regulating and promoting bottle reuse, then such mega-retailers can extinguish the practice with relative ease. This outcome serves neither customers nor the environment, but instead benefits producers and sellers.[33]

Beyond Waste

Another approach gaining momentum among environmentalists over the past decade is called "zero waste." Zero waste refers to eliminating refuse before it gets made, at the front end, instead of the current norm of treating trash only after it already exists, at the back end. Some advocates explain that zero waste sprang from the inadequacies of recycling, since the latter does not directly curb waste production.[34] This new method is more comprehensive than reprocessing alone: "Zero waste maximizes recycling, minimizes waste, reduces consumption and ensures that products are made to be reused, repaired or recycled back into nature or the marketplace."[35] Similar to green capitalism's program to redesign industrial production, and a further development of past reuse and recycling practices like 1970s source reduction, zero waste works through grassroots activism and policy-level advocacy to foster deep structural reforms.

According to the GrassRoots Recycling Network, zero waste centers on "corporate responsibility for wastes, government policies for resource conservation, and sustainable jobs from discards."[36] Although they share many aims, zero waste differs from green capitalism in that the state plays a key role in regulating discard levels and natural resources. Zero-waste proponents Bill Sheehan and Daniel Knapp explain that government intervention is warranted if asking business to change is not effective. Federal, state and local officials should step in to "change rules and laws to reward resource conserving behavior and penalize resource wasting behavior."[37]

Enforcing production modifications is an important aspect of zero waste because so much refuse is the result of commodity packaging and built-in obsolescence, and most castoffs are actually created during manufacturing. A zero-waste system would require manufacturers to use nontoxic biodegradable materials that could be safely returned to the earth; maximum recirculation would be mandatory for any hazardous nonrenewable inputs that were absolutely necessary. According to some no-wasters, whatever does not fit into those two categories should be banned. Zero waste thus aims to retool industry so that it relies only on resources that are easily reusable.[38]

This anti-trash system also seeks to incorporate environmental costs into the price of a commodity, ending the long-time industry practice of externalizing these costs. As a result, that bottle of water for sale at the corner store would carry a higher price tag and might even list ingredients that more accurately reflect the product's contents—like waste from drilling for the natural gas used to make resin for the container.

But these changes are impossible according to pro-waste lobbyists and their industry clients. Those opposed to transformations like zero waste use the old chimera of mass layoffs and ensuing poverty. This assertion is as baseless today as it was in the 1970s when the same bogeyman was enlisted against beverage container deposit laws. Arguments that reusing discards would eliminate manufacturing jobs are vague and as yet unproven, while recycling actually increases the demand for labor, creating ten times more jobs per ton than landfilling or incineration.[39] What's more, per-ton, recycling-based manufacturers can employ up to sixty times more workers than do landfills.[40] In 2000, North Carolina had almost 9,000 employees in the recycling trades while a much smaller proportion were displaced from old jobs—ninety waste handlers and three timber harvesters.[41] Zero waste promises to generate opportunities for small businesses, boost the need for skilled labor, and enrich local economies rather than siphon off revenues and jobs.[42]

Zero-waste advocates have the goal of shutting down landfills and incinerators altogether. While this might sound implausible, it is important to note there are still cultures that have yet to formulate a word for garbage because it is incomprehensible that any object could be useless.[43] Zero-waste programs have already been implemented in places like Toronto; Canberra Territory, Australia;

and Halifax, Nova Scotia; and almost half of New Zealand's local authorities have committed to eliminate landfilled waste by 2015.[44] Zero waste poses the idea of garbage as an option, not some inevitable outcome of a natural system. In other words, discarded materials do not have to be wasted.

Garbage In, Garbage Out

To truly confront the issue of garbage and its impact on the environment two fundamental shifts must take place. First, trash needs to be addressed in terms of production instead of consumption. Pinning responsibility for the ever-growing swells of packaging and broken, outmoded commodities on individual consumers has only facilitated manufacturing's expanded and intensified wasting. Additionally, while consumers making choices with the environment in mind is a good thing, it is in no way a real solution to our trash woes. Second, industry's inability to regulate itself must be acknowledged and replaced with enforceable environmental measures. Failing to remake production and sidestepping legislatively mandated protections will only bring further environmental degradation.

In encouraging waste reduction before garbage gets produced, Germany's system and zero-waste programs do not operate just in the spheres of circulation and consumption, they begin to reach back into the realm of production. This profoundly distinguishes them from other strategies that focus on the acts of the consumer and aim to manage wastes only after they're already made. Extended producer responsibility and zero waste programs broach a more active role for the public as stewards of human and ecological health by influencing the production process. However, in the United States and many other capitalist countries, manufacturing has remained largely untouched, allowing business to operate on its own terms, in what Marx called "the hidden abode of production."[45]

Over the last thirty years, when individual towns, states and countries have passed laws restricting the disposal of certain materials—closing a dump site or prohibiting toxic substances—without limiting the *creation* of those wastes, companies have continued making the stuff. And when the resulting trash can't be treated locally, it just gets exported elsewhere. Likewise, when recycling and recycled content in commodities is not mandated, U.S. firms have shown they will not implement the practice in a meaningful way. Anything short of government enforcement of production regulations to protect human and environmental health will ultimately end in failure.

And if these measures are to successfully govern and protect natural systems, then the state must act in the public interest and not as an agent of business. It is not enough simply to turn over regulatory control to governments without addressing the injustice so often inherent in the state. Currently, the EU is pursuing a legal case against container laws in Denmark and Germany, arguing that these countries' packaging regulations violate free trade. Likewise, U.S. courts consistently support the export of garbage on the grounds that restricting it would illegally hinder interstate commerce. Putting the well-being of commerce above the health of the environment that everyone depends on reveals a profound lack of real democracy. As sociologist Joel Kovel writes, "The struggle for an ecologically rational world must include a struggle for the state, and since the state is the repository of many democratic hopes, it is a struggle for the *democratization* of the state."[46] There must be democratic decision-making in the regulation of industry and the use of natural resources.

When the public has no say in manufacturing—which materials get used; how they are extracted from nature; what kind of production process is employed; and levels of toxicity in manufacturing, use, and disposal of materials—the democratic utilization of our common natural resources is fundamentally undermined. Just as the airwaves belong to the people, so too is the rest of the natural world part of the public commons. It is only fair that since we share the responsibility for and consequences of wastes, we should also participate in the choices made during production, the real source of trash. U.S. industry's largely unfettered access to natural resources is the mark of a deeply undemocratic system.[47]

Industry and government justify ever-more disposable commodities, growing mountains of trash, and environmental destruction from unchecked production as necessary for a healthy economy that provides jobs and a high standard of living. However, according to the U.S. Census, the country has grown more economically polarized over the last thirty years. Today there is greater income inequality than any time since World War II. This disparity increased throughout the 1990s as the middle class shrank even in the midst of the longest period of economic expansion in fifty years.[48] And, as of 2000, the poorest 20 percent of households received just 3.6 percent of the country's total household income, while the richest 20 percent took home 49.7 percent.[49] Today, layoff rates are higher than they have been since the deep recession of the early 1980s.[50] Under the current U.S. market system, since the year 2000 more than 2 million jobs have left the country due to outsourcing, replaced by half as many low-wage, unprotected service-sector jobs.

This means that all those trashed appliances, cars, clothes, and the mountains of wasted packaging are actually not the product of an economy that delivers its benefits to the most people. On the contrary, the biggest beneficiaries of a trash-rich marketplace are those at the top. Garbage is the detritus of a system that unscrupulously exploits not only nature, but also human life and labor. Why should Americans risk their health and the survival of natural systems to enrich the nation's elite? Though the benefits of trash are unequally distributed, pollution threatens the natural systems that affect everyone.

Since the 1970s predictions of an environmental apocalypse have abounded, but today's supply of food, manufactured goods, fossil fuels and clean water seems to indicate that the natural world is just fine. This is because in the market economy deeper environmental destruction is kept hidden, cloaked by the commodity form. Since consumers find the finished product in the store and don't see the piles of mining slag, clear-cut forests, and air and water pollution resulting from the item's production and disposal, they more easily believe that waste is manageable, while the devastation of nature remains an abstraction. This destruction is further obscured by the pro-waste bloc, which consistently generates messages minimizing the effects of industrial production on nature. As sociologist Leslie Sklair argues, these interests aim to "deflect attention from the idea of a singular ecological crisis and to build up the credibility of the idea that what we face is a series of manageable environmental problems."[51] But, in reality, ecological cataclysm has been unfolding unevenly across the globe over the last several decades.

Garbage, the miniature version of production's destructive aftermath, inevitably ends up in each person's hands, and it is proof that all is not well. Trash therefore has the power to unmask the exploitation of nature that is crystallized in all commodities. Garbage reveals the market's relation to nature; it teases out the environmental politics hidden inside manufactured goods. Because of this, transforming the way our society conceives of and treats the everyday substance of garbage would have profound effects in other areas of ecological crisis, such as dying oceans, ozone depletion, global warming, and the proliferation of toxic chemicals throughout our food, water and air.

Notes

Introduction

1. Parts of this introduction were adapted from an essay written collaboratively with Christian Parenti. Fresh Kills was opened in 1948 by Robert Moses as a temporary dump site. See Benjamin Miller, *Fat of the Land: Garbage of New York—The Last Two Hundred Years* (New York: Four Walls Eight Windows, 2000), pp. 198–212.

2. That's 236 million tons of municipal solid waste. See EPA website, http://www.epa.gov/epaoswer/non-hw/muncpl/facts.htm. Other studies put the figure closer to 369 million tons of garbage yearly, or 7 pounds per person per day. See Scott Kaufman, "National Garbage Survey Highlights Opportunities for Americans to Move from Being Waste-Full to Waste-Wise," *Earth Institute News* (http://www.earthinstitute.columbia.edu/news/2004/story01-23-04.html).

3. U.S. EPA website: see http://www.epa.gov/epaoswer/non-hw/muncpl/facts.htm.

4. U.S. EPA, Office of Solid Waste and Emergency Response, *Municipal Solid Waste in the United States: 2001 Facts and Figures* (Washington, D.C., 2003), pp. 3–4.

5. Paige Wiser, "Curbing Enthusiasm for Consumerism," *Chicago Sun-Times,* Oct. 9, 2003.

6. U.S. EPA, "Frequent Questions on Landfill Gas and How It Affects Public Health, Safety and the Environment," see www.epa.gov/lmop/faq-3.htm#2.

7. GrassRoots Recycling Network, *Wasting and Recycling in the United States 2000* (Athens, Ga., 2000), pp. 17, 30; Neil Seldman, "Recycling—History in the United States," in *Encyclopedia of Energy Technology and the Environment,* ed. Attilio Bisio and Sharon Boots (Hoboken, N.J.: John Wiley, 1995), p. 2359.

8. On the portion of packaging made from plastics, see Ecology Center, *Report of the Berkeley Plastics Task Force* (Berkeley, Calif., April 8, 1996), p. 4. On the quantity of packaging in landfills, see U.S. EPA, *Municipal Solid Waste,* p. 7.

9. For the estimate of 1,000 years, see Brian Howard, "Message in a Bottle," *E: The Environmental Magazine* 14, no. 5 (Sept./Oct. 2003), p. 36.

10. William McDonough and Michael Braungart, *Cradle to Cradle: Remaking the Way We Make Things* (New York: North Point Press, 2002).

11. Maraline Kubik, "Northeast Ohio Leads U.S. in Plastics Production," *Business Journal of the Five-County Region* 13, no. 14, (March 1, 1997).

12. Paul Goettlich, "The Sixth Basic Food Group," Nov. 16, 2003, see www.mindfully.org/Plastic/6th-Basic-Food-Group3.htm. The amount of debris found around Antarctica increased 100-fold in the early 1990s; much of this new seaborne refuse is plastic and carries sea creatures with the tides to new parts of the globe, seriously threatening biodiversity. See Hillary Mayell, "Ocean Litter Gives Alien Species an Easy Ride," *National Geographic News,* April 29, 2002.

13. Seldman, "Recycling," p. 2352.

14. On paper accounting for half the trash in U.S. landfills, see "The People vs. The People," *Colors* 40 (2001), p. 27.

15. Municipal solid waste numbers are from the Environmental Protection Agency, see www.epa.gov/epaoswer/non-hw/muncpl/facts.htm.

16. On global warming's increase, see Maggie Fox, "Global Warming Effects Faster Than Feared—Experts," Reuters, Oct. 21, 2004. On U.S. carbon emissions, see "Americans as Consumers of and Contributors to World Resources," *Research Alert,* July 5, 2002.

17. On Caribbean storms, see Mike Toner, "Fury of Storms Linked to Warming," *Atlanta Journal-Constitution,* Oct. 22, 2004. On Bangladesh, see Lucy Ward, "Bangladesh Suffers in Silence," *Guardian Weekly,* Oct. 8–14, 2004.

18. Sushi Das, "Dirty Old Bags," *The Age,* June 29, 2004.

19. Jon E. Hilsenrath, "Beijing Strikes Gold with U.S. Recycling," *Asian Wall Street Journal,* April 9, 2003.

20. On the amount of consumer spending, see Steve Lohr, "Maybe It's Not All Your Fault," *New York Times,* Dec. 5, 2004. In 1996, spending for municipal solid waste—household garbage—was $43.5 billion; see GrassRoots Recycling Network, *Wasting and Recycling in the United States 2000*, p. 4. More recent reports put that number closer to $70 billion today. See Neil Seldman, "The New U.S. Recycling Movement," paper presented at the Michigan Recycling Coalition Annual Conference, May 2004, p. 2.

Chapter 1

1. The research for this section was conducted in 2002 at Norcal's San Francisco Recycling & Disposal, Inc., facility.

2. Since the research for this book was done, Norcal has opened a new plant for sorting recyclable materials called Recycle Central, which is located at a different site. The facility I visited is still the city's main transfer station and it still processes some recyclables. And although the procedures at Recycle Central differ in some ways, the processes at Norcal's MRF are typical of transfer stations throughout the United States and therefore warrant inclusion.

3. Eric Lipton, "As Imported Garbage Piles Up, So Do Worries," *Washington Post*, Nov. 12, 1998.

4. The research for this section was conducted in November 2004 and is based on a tour and interview with Robert Iuliucci, Waste Management Inc. district manager, and Geri Rush, community relations coordinator at WMI's GROWS and Tullytown, Penn., landfills. Additional research on landfills is based on interviews with Thomas McMonigle, landfill engineer at Chrin Brothers, Inc., in Easton, Penn.

5. As of 2003, New York and New Jersey "were by far the biggest out-of-state contributors to Pennsylvania's landfills." See Marc Levy, "PA Trash Imports Decline for First Time in 11 Years," Associated Press, June 4, 2003.

6. Lipton, "As Imported Garbage Piles Up."

7. As of 1998, BFI had pled guilty to violating the Clean Water Act in Virginia and was fined for its violations but, despite recurrent illegal cocktailing, WMI had yet to be disciplined. See Lipton, "As Imported Garbage Piles Up."

8. *Comments in Opposition to Proposed Rule Deregulating Municipal Solid Waste Landfills*, Grassroots Recycling Network, Natural Resources Defense Council, and Friends of the Earth, Aug. 9, 2002. See www.grrn.org/landfill/landfill_summary.html.

9. GrassRoots Recycling Network, *Wasting and Recycling*, p. 16.

10. The research for this section was conducted in November 2004 and is based on a tour and interview with Narayan Dave, environmental engineer at American Ref-Fuel's Newark, N.J., incinerator. I was told about this facility by garbage anthropologist Robin Nagle .

11. From a November 2004 interview with Thomas McMonigle.

Chapter 2

1. Catharine E. Beecher, *Treatise on Domestic Economy* (New York: Marsh, Capen, Lyon and Webb, 1841), p. 373. This book was as much an ideological tract that

sought to define the new role of middle-class women as it was a useful manual for homemakers.

2. Susan Strasser, *Waste and Want: A Social History of Trash* (New York: Henry Holt, 1999), p. 26.

3. Alan Taylor, *American Colonies (The Penguin History of the United States)*, edited by Eric Foner (New York: Penguin, 2001), p. 311.

4. On littered colonial farms, see Gail Collins, *America's Women: 400 Years of Dolls, Drudges, Helpmates and Heroines* (New York: Morrow, 2002) p. 62. The harm wrought by unchecked garbage dumping came in the form of disease, discussed in chapter 3.

5. Taylor, *American Colonies*, p. 146.

6. See Collins, *America's Women*, p. 62.

7. J. Buel, "Remarks on the Construction and Management of Cattle Yards," *American Farmer* 8, no. 16 (July 7, 1826), p. 122.

8. Richard A. Wines, *Fertilizer in America: From Waste Recycling to Resource Exploitation* (Philadelphia: Temple University Press, 1985), p. 6. Between 1790 and 1840 the number of cities in the United States grew from 24 to 131. See U.S. Bureau of the Census, *Characteristics of the Population*, in vol. 1 of *Census of Population: 1960*, pt. A, pp. 1-14, 1-15, table 8.

9. Wines, *Fertilizer in America*, p. 154.

10. *American Farmer* 14, no. 43 (1833), p. 339.

11. Marc Linder and Lawrence S. Zacharias, *Of Cabbages and Kings County: Agriculture and the Formation of Modern Brooklyn* (Iowa City: University of Iowa Press, 1999), p. 46.

12. *Working Farmer*, no. 3 (1851), p. 148.

13. Edwin G. Burrows and Mike Wallace, *Gotham: A History of New York City to 1898* (New York: Oxford University Press, 1999), p. 787; Linder and Zacharias, *Of Cabbages and Kings County*, pp. 49–50.

14. Refuse traders hauled animal dung "to the outskirts of the city where they composted it with sawdust, spent tanner's bark, spent charcoal from the rectifying establishments, and other urban wastes to produce a light, friable manure for which farmers were willing to pay premium prices." See Wines, *Fertilizer in America*, pp. 9–10, 11.

15. Ibid., p. 11.

16. Ibid., pp. 12–13.

17. Karl Marx, *Capital, Volume I* (London: Penguin, 1990), trans. Ben Fowkes, p. 637.

18. This was reported in New York City newspapers of the day. See Gert H. Brieger, "Sanitary Reform in New York City: Stephen Smith and the Passage of the Metropolitan Health Bill," in *Sickness and Health in America: Readings in the History of Medicine and Public Health*, ed. Judith Walzer Leavitt and Ronald L. Numbers (Madison: University of Wisconsin Press: 1985), p. 405.
19. Benjamin Miller, *Fat of the Land: Garbage of New York—The Last Two Hundred Years* (New York: Four Walls Eight Windows, 2000), p. 40.
20. This practice continued, although on a much smaller scale, well into the 1950s. See Strasser, *Waste and Want*, pp. 25–26.
21. Ibid., pp. 69–71.
22. Jacob A. Riis, *How the Other Half Lives* (New York: Penguin, 1997 [1890]), p. 42.
23. Colonel A.H. Rogers, former deputy commissioner of New York City's Street Cleaning Department, quoted in Miller, *Fat of the Land*, p. 76.
24. James D. McCabe, Jr., *New York by Gaslight* (New York: Arlington House, 1984 [1882]), pp. 584–85.
25. *Report of the Council of Hygiene and Public Health of the Citizen's Association of New York upon the Sanitary Condition of the City* (1863), cited in Miller, *Fat of the Land*, p. 66.
26. Cited in ibid., p. 78.
27. Brieger, "Sanitary Reform in New York City," p. 402. According to one source at the time, mortality rates among tenement-dwelling New Yorkers were seven times the city's average overall rate. See "Tenement Life in New York," *Harper's Weekly* 23 (1879), p. 246. Also see Riis, *How the Other Half Lives*, p. 31.
28. Quoted in Herbert Asbury, *The Gangs of New York: An Informal History of the Underworld* (New York: Thunder's Mouth Press, 1998 [1927]), p. 11.
29. Riis, *How the Other Half Lives*, pp. 9–10.
30. Gray Brechin, author of *Imperial San Francisco* (Berkeley: University of California Press, 1999), has made a comparison between that city's skyscrapers and the Sierra mines they were financially dependent upon.
31. New York Association for the Improvement of the Condition of the Poor, *First Report of a Committee on the Sanitary Condition of the Laboring Classes in the City Oe* [sic] *New York* (New York: John F. Trow, 1853).
32. Hogs ran through the streets of cities across the country, like Washington, D.C., where they roamed freely, as did rats, which infested the White House. See Constance Green, *Washington: Village and Capital, 1800–1878* (Princeton: Princeton University Press, 1962). On New York's 1849 pig crackdown, see Charles E. Rosenberg, *The Cholera Years* (Chicago: University of Chicago Press, 1962), p. 113.

33. Hendrick Hartog, "Pigs and Positivism," *Wisconsin Law Review* (July/Aug. 1985), electronic version.
34. Rosenberg, *Cholera Years*, p. 113.
35. Ted Steinberg, "Down to Earth," *American Historical Review* 107, no. 3 (June 2002), p. 811; Burrows and Wallace, *Gotham*, p. 786; Miller, *Fat of the Land*, p. 36.
36. Rosenberg, *Cholera Years*, p. 114.
37. New York Association for the Improvement of the Condition of the Poor, *First Report*, p. 4. The AICP described a typical tenement in this way: "The pestiferous stench and filth of these pent-up tenements exceed description. 'In one room,' says a Visitor, 'six people are living, with hens scratching about on the bed'" (p. 12). In fact, episodic boards and commissions of health were common, if ineffectual and short-lived, throughout the eighteenth and nineteenth centuries, arising in response to disasters and epidemics. In New York, for example, the state-appointed Health Commission was formed in the aftermath of the 1793 onslaught of yellow fever. Threat of further pestilence spurred the 1803 creation of the City's Board of Health. Such bodies usually responded to disease with time-tested measures like quarantining infected homes and sailing vessels. See Burrows and Wallace, *Gotham*, p. 358; Rosenberg, *Cholera Years*, p. 19.
38. *Edinburgh Review*, quoted in Miller, *Fat of the Land*, p. 65. This British publication was widely read by the American middle classes on the East Coast. For more on the connection between morality and physical cleanliness, see New York Association for the Improvement of the Condition of the Poor, *First Report*.
39. T.L. Cuyler, quoted in Charles E. Rosenberg and Carroll Smith-Rosenberg, "Pietism and the Origins of the American Public Health Movement: A Note to John H. Griscom and Robert M. Hartley," in Leavitt and Numbers, *Sickness and Health in America*, p. 397n34.
40. "Article VII—The United States Sanitary Commission. A Sketch of Its Purposes and Its Work. Compiled from Documents and Private Papers," *North American Review* 98, no. 202 (Jan. 1864), p. 163; "Article VIII," *North American Review* 98, no. 203 (April 1864). The U.S. Sanitary Commission (USSC) was the brainchild of groups like the Ladies' Union Aid Society and the Women's Central Association, and fashioned after the British Sanitary Commission, which was an outgrowth of Florence Nightingale's work in the Crimean War. Two prominent women in the USSC, Josephine Shaw Lowell and Louisa Lee Schuyler, went on to the State Charities Aid Association to monitor the unsanitary conditions and moral pollution of New York's underside. See Burrows and Wallace, *Gotham*, p. 1031.
41. Two and a half years earlier New York had halted street cleaning outright when

the department's budget was depleted, so it is likely that no systematic cleaning was in place by 1863. See Brieger, "Sanitary Reform in New York City," p. 403. On the connection between sanitation and riots, see Miller, *Fat of the Land*, p. 64.

42. "The Draft," *New York Times*, July 13, 1863.

43. On the Draft Riot's racial violence, see Leslie M. Harris, *In the Shadow of Slavery: African Americans in New York City, 1626–1863* (Chicago: University of Chicago Press, 2003), pp. 279–88.

44. "The Mob in New York," *New York Times*, July 14, 1863; "The Riot in Second Avenue," *New York Times*, July 15, 1863.

45. "The Mob in New York."

46. Ibid.; "The Riot in Second Avenue."

47. "Facts and Incidents of the Riot," *New York Times*, July 16, 1863.

48. "The Mob in New York"; "The Reign of the Rabble," *New York Times*, July 15, 1863; "Destruction of Grain-Elevator in the Atlantic Dock by Fire—Great Destruction of Property," *New York Times*, July 16, 1863; "Facts and Incidents of the Riot."

49. "The Mob in New York."

50. "Facts and Incidents of the Riot;" "The Reign of the Rabble." For more on citizens' volunteer militias, see "Another Day of Rioting," *New York Times*, July 16, 1863; "Excitement in Jamaica—A Number of Stores Robbed," *New York Times*, July 16, 1863.

51. On troops being called back from war service, see, "Another Day of Rioting."

52. Burrows and Wallace, *Gotham*, pp. 892–95.

53. "Tenement Houses—Their Wrongs," *New York Tribune*, Nov. 23, 1864, p. 4.

54. Charles Loring Brace, *The Dangerous Classes of New York and Twenty Years among Them* (New York: Wyncoop & Hallenbeck, 1872), p. 31.

55. This New York organization was modeled on similar British organizations like the Health of Towns Association. On the Health of Towns Association's influence on the Citizens' Association, see Brieger, "Sanitary Reform in New York City," p. 406.

56. Burrows and Wallace, *Gotham*, p. 784; Miller, *Fat of the Land*, pp. 20–21; Rosenberg and Smith-Rosenberg, "Pietism," p. 385.

57. Rosenberg, *Cholera Years*, p. 112.

58. Miller, *Fat of the Land*, p. 44.

59. George E. Hooker, "Cleaning Streets by Contract—A Sidelight from Chicago," *Review of Reviews* 15, no. 3 (March 1897), pp. 439–40.

60. "Cincinnati's Dirty Streets," *New York Times*, Nov. 23, 1882.

61. Brieger, "Sanitary Reform in New York City," p. 405; Miller, *Fat of the Land*, pp.

64–65. The surveillance methods developed by these early sanitarians, so thorough and accurate, were later adopted by the federal Census Bureau.
62. On the epidemiological milestone, see Brieger, "Sanitary Reform in New York City," p. 405. On the previous work of missionaries, see Rosenberg and Smith-Rosenberg, "Pietism," p. 390.
63. *Report of the Council of Hygiene*, quoted in Miller, *Fat of the Land*, p. 65. Also see Brieger, "Sanitary Reform in New York City," p. 405.
64. "An Act to Create a Metropolitan Sanitary District and Board of Health Therein for the Preservation of Life and Health and to Prevent the Spread of Disease" was the law's formal name; it passed on Feb. 26, 1866. See Brieger, "Sanitary Reform in New York City," pp. 400–401, 407–8.
65. Richard L. Bushman and Claudia L. Bushman, "The Early History of Cleanliness in America," *Journal of American History* 74 (March 1988), pp. 1234–36.
66. Catharine E. Beecher quoted in Strasser, *Waste and Want*, p. 32.
67. Even before the Civil War, home looms were increasingly unable to compete with factory-produced cloth; the cost of plain shirting fell from 42¢ a yard in 1815 to 7.5¢ in 1830, and prices continued to drop in the postwar decades. See Rolla Milton Tryon, *Household Manufactures in the United States, 1640–1860: A Study in Industrial History* (Chicago: University of Chicago Press, 1917), p. 276.
68. Caroline L. Hunt, "Home Problems from a New Standpoint: More Pleasure for the Producer of Household Stuff," *Chautauquan* 37, no. 2 (May 1903), p. 178.
69. On pre-1850 figures, see Miller, *Fat of the Land*, p. 19. On 1850–90 figures, see Ira Rosenwaike, *Population History of New York City* (Syracuse: Syracuse University Press, 1972), pp. 42, 67. On the 1900 figure, see Kenneth T. Jackson, *The Encyclopedia of New York City* (New Haven: Yale University Press and The New-York Historical Society, 1995), p. 921.
70. Dominique Laporte, *History of Shit* (Cambridge: MIT Press, 2002), p. 39.
71. Martin V. Melosi, *Garbage in the Cities: Refuse, Reform and the Environment 1880–1980* (Chicago: Dorsey Press, 1981), p. 35; Strasser, *Waste and Want*, pp. 121–22; on New York see, "The Citizens' Committee," *New York Times*, April 9, 1881.
72. Melosi, *Garbage in the Cities*, pp. 114, 122–23.
73. Quoted in ibid., p. 36.
74. Maureen Ogle, *All the Modern Conveniences: American Household Plumbing, 1840–1890* (Baltimore: Johns Hopkins University Press, 1996), p. 102. For more on sanitary science, see John S. Billings, M.D., "The World's Debt to Sanitary Science," *Chautauquan* 21 (1895).
75. Melosi, *Garbage in the Cities*, pp. 56, 59–60.

76. Ibid., pp. 59, 69.
77. Ibid., pp. 53, 65–66, 74–75.
78. George E. Waring, Jr., quoted in Daniel Thoreau Sicular, "Currents in the Waste Stream" (master's thesis, University of California, Berkeley, 1981), p. 33.
79. George E. Waring, Jr., quoted in Strasser, *Waste and Want*, p. 140.
80. George E. Waring, Jr., *Street Cleaning and Its Effects* (New York: Doubleday and McClure, 1898), p. 23; "The Disposal of New York's Refuse," *Scientific American* 89, no. 17 (Oct. 24, 1903), p. 292.
81. "The Utilization of New York City Garbage," *Scientific American* 78, no. 7 (Aug. 14, 1897), p. 102. On the culture of the island, see Miller, *Fat of the Land*, pp. 85, 88.
82. "The Disposal of New York's Refuse,"*Scientific American*, p. 292. Also see Rudolph Hering and Samuel A. Greeley, *Collection and Disposal of Municipal Refuse* (New York: McGraw Hill, 1921), pp. 298–99.
83. Hering and Greeley, *Municipal Refuse*, p. 444. The twentieth-century Dada artist Kurt Schwitters, who used castoff materials to make collages and architectural sculptures, is often mistakenly credited with coining the term *Merz*. Schwitters used the word as both a noun and a verb to refer to all aspects of his art, even his poems.
84. By 1899 twenty reduction plants had been constructed and, by 1914, forty-five had been built; however, not all remained open and functioning. See Sicular, "Currents in the Waste Stream," pp. 49, 52. On plant locations, see Hering and Greeley, *Municipal Refuse*, p. 448.
85. "The Utilization of New York City Garbage," p. 102.
86. At its peak, around the turn of the nineteenth century, Barren Island's factories received 3,000 tons of household and commercial garbage per day along with all the carcasses and offal from Manhattan, Brooklyn, and the Bronx. From this mammoth flow of waste was rendered a yearly take of 50,000 tons of oils, and tens of thousands of tons of grease, fertilizers and other by-products worth more than $10 million. See Miller, *Fat of the Land*, pp. 44, 85.
87. Refuse sorting and other forms of salvaging were practiced on the island before the introduction of "reduction" in the 1880s. On Moses evicting the islanders, see Miller, *Fat of the Land*, p. 191.
88. William F. Morse quoted in Sicular, "Currents in the Waste Stream," p. 46.
89. G.E. Waring, "The Utilization of City Garbage," *Cosmopolitan Magazine* 24 (Feb. 1898), pp. 406–8. In this thinking Waring was not alone. Chicago's Health Commissioner, Dr. Arthur R. Reynolds, agreed about treatment in his own municipality: "While much of the city's waste is combustible and can only be disposed of by fire,

I believe it an economical error to burn up garbage, for the reason that it will yield an item of considerable value by extracting the grease and fertilizer from it." See Dr. Arthur R. Reynolds, quoted in "Some Financial, Political and Sanitary Phases of Garbage Disposal," *Engineering News-Record* 45, no. 7 (Feb. 14, 1901), p. 120.

90. Macdonough Craven, quoted in Sicular, "Currents in the Waste Stream," p. 70.

91. "The Utilization of New York City Garbage," p. 102.

92. William F. Morse, "The Disposal of the City's Waste," *American City* 2, no. 4 (April 1910), p. 180.

93. From a house card "Regulations for Collection of Household Waste," distributed by the Springfield, Mass., Department of Streets and Engineering, reproduced in Hering and Greeley, *Municipal Refuse*, p. 98.

94. "Disposal of New York's Refuse,"p. 292; Melosi, *Garbage in the Cities*, p. 51.

95. Melosi, *Garbage in the Cities*, p. 100.

Chapter 3

1. "Doctors recognized the advantages of eradicating filth and cleaning the physical surroundings, but they had been frustrated for years by their inability to prevent communicable diseases through sanitary measures alone." See Melosi, *Garbage in the Cities,* p. 80.

2. During the 1832 cholera outbreak in New York City, locals drained their bank accounts and fled the city, shutting down commerce. With the city largely abandoned, some who stayed behind took to looting deserted shops, houses and offices; and some stole documents in order to commit identity theft and fraud. The spatiality of epidemic disease was indeed bad for business. See Rosenberg, *Cholera Years*, pp. 30–33.

3. Billings, "Sanitary Science," p. 22.

4. A 1925 study by the American Child Health Association revealed that only 10 percent of the cities surveyed still charged their health departments with waste management. See Melosi, *Garbage in the Cities*, pp. 80–84. For more on this separation, see George A. Soper, "The First International Conference on Public Cleansing," *Municipal Sanitation* 2, no. 12 (Dec. 1931), p. 582.

5. Rudolph Hering, quoted in Sicular, "Currents in the Waste Stream," p. 12.

6. Between 1899 and 1905 about 85 percent of U.S. cities with populations exceeding 25,000 began using mechanical street sweeping equipment. See Melosi, *Garbage in the Cities*, pp. 141–42.

7. On the various uses of streets in the early twentieth century, see Carol Aro-

novici, "Municipal Street Cleaning and Its Problems," *National Municipal Review* 1, no. 2 (April 1912), pp. 218–19.

8. A.L. Thompson, "A Review of British Practices in Street Sanitation," *Municipal Sanitation* 2, no. 2 (Feb. 1931), p. 62.

9. Richard E. Fogelsong, *Planning the Capitalist City: The Colonial Era to the 1920s* (Princeton: Princeton University Press, 1986), p. 200.

10. Melosi, *Garbage in the Cities*, p. 137. This shift in the use of streets was solidly established by the 1930s: "The term 'street' used upon a map of a town or city, imports a public way for the free passage of its trade and commerce. It is of that character that, if a person willfully obstructs it, he can be prosecuted under the penal laws and be made responsible for the expenses of removing the obstruction." See Leo T. Parker, "Sewer Construction and the Law," *Municipal Sanitation* 2, no. 2 (Feb. 1931), p. 82. Also see Edward T. Hartman, "The Social Significance of Clean Streets," *American City* 3, no. 10 (Oct. 1910), p. 173.

11. Henri Lefebvre, *The Urban Revolution* (Minneapolis: University of Minnesota Press, 2003), p. 20.

12. Susan Strasser, *Waste and Want: A Social History of Trash* (New York: Henry Holt, 1999), p. 120.

13. Melosi, *Garbage in the Cities*, p. 151.

14. Ibid., p. 137.

15. Ibid., p. 139.

16. "Street Cleaning in Philadelphia," *Municipal Journal* 40, no. 26 (June 29, 1916), p. 898.

17. W.S. Webb, "Garbage Collection by Contract—Costs: Contract Collection and Incineration at Houston, Texas," *American City* 50, no. 10 (Oct. 1935), p. 15.

18. "Street Cleaning Standards," *Municipal Journal and Public Works* 35, no. 24 (Dec. 11, 1913), p. 795.

19. On the number of horses, see Melosi, *Garbage in the Cities*, p. 25. On the increase in automobile traffic, see "Street Cleaning Standards," p. 795.

20. K.L. Rothermund, "Ohio Starts Revamping Roads," *Engineering News-Record* 120, no. 11 (March 17, 1938), p. 411.

21. Melosi, *Garbage in the Cities*, p. 25.

22. Rosalyn Baxandall and Elizabeth Ewen, *Picture Windows: How the Suburbs Happened* (New York: Basic Books, 2000), p. 17.

23. Between 1903 and 1918, overall U.S. waste output ranged from 1,000 to 2,000 pounds per capita, with ashes comprising the bulk at anywhere from 300 to 1,500 pounds; garbage (food and organic wastes) made up 100 to 300 pounds; and rub-

bish (all other discards) accounted for 25 to 125 pounds. See Melosi, *Garbage in the Cities*, p. 160.

24. Ibid., pp. 160–61.

25. "Salvaging Municipal Refuse in Three Cities," *Public Works* 62, no. 2 (Feb. 1931), pp. 23–24.

26. Hering and Greeley, *Municipal Refuse*, p. 29.

27. Melosi, *Garbage in the Cities*, p. 190.

28. Strasser, *Waste and Want*, pp. 175–78.

29. Disposable sanitary napkins had been on the market since the end of the nineteenth century, but "widespread use of disposable pads did not begin until Kimberly-Clark introduced Kotex in 1920." See Strasser, *Waste and Want*, pp. 162–63.

30. On such packaging, see Strasser, *Waste and Want*, pp. 171–72.

31. Lauren R. Hartman, "Forever Flexible," *Packaging Digest* 36, no. 12 (Nov. 1, 1999).

32. Of the 146 cities surveyed, "almost all of them dumped ash; 80 dumped rubbish; and over 90 dumped their garbage, mostly on land, though some in water." See Hering and Greeley, *Municipal Refuse*, pp. 240–57.

33. Melosi, *Garbage in the Cities*, p. 167. Yellowstone National Park used to have bleacher seats at its dump from which tourists could watch the foraging "dump bears."

34. J.H. Hildreth, quoted in Miller, *Fat of the Land*, p. 79.

35. Hering and Greeley, *Municipal Refuse*, p. 241.

36. "A Five Year Plan on Sanitary Fills," *Engineering News-Record* 123, no. 5 (Aug. 3, 1939), p. 65.

37. Chicago Department of Public Works 1882 annual report, quoted in Melosi, *Garbage in the Cities*, p. 42.

38. Louisiana Board of Health 1898 annual report, quoted in Melosi, *Garbage in the Cities*, p. 166.

39. On Milwaukee and other towns that used water dumping, see Hering and Greeley, *Municipal Refuse*, p. 241. On contamination in general, see Kevin Lynch, *Wasting Away, An Exploration of Waste* (San Francisco: Sierra Club Books, 1990), p.

40. Miller, *Fat of the Land*, p. 71.

41. The event occurred in 1916. See C.G. Gillespie and E.A. Reinke, "Municipal Refuse Problems and Procedures," *Civil Engineering* 4, no. 9 (Sept. 1934).

42. Quoted in Melosi, *Garbage in the Cities*, p. 47. Also see Roger J. Bounds, "A Survey of Practices in Refuse Disposal in American Cities," *Municipal Sanitation* 2, no. 9 (Sept. 1931), p. 435.

43. Sicular, "Currents in the Waste Stream," p. 42; Melosi, *Garbage in the Cities*, p. 48.
44. Charles Gilman Hyde, "Sanitary Engineering as a Vocation," *Municipal Sanitation* 5, no. 5 (May 1934), p. 155.
45. Ogle, *All the Modern Conveniences*, p. 102.
46. Hyde, "Sanitary Engineering as a Vocation," p. 155.
47. Edwin T. Layton, Jr., *The Revolt of the Engineers: Social Responsibility and the American Engineering Profession* (Baltimore: Johns Hopkins University Press, 1986), pp. 2–3.
48. A more recent example of this: In 1972 three engineers working on Bechtel's construction of the Bay Area Rapid Transit commuter train in California blew the whistle on defects in the control system. They were castigated by fellow engineers as unethical and their careers were suffocated. The claims the three engineers made were later proven correct after accidents led to a public investigation. See Layton, *Revolt of the Engineers*, pp. xi–xii.
49. Layton, *Revolt of the Engineers*, p. 208. In 1925 the American Engineering Council rejected the government's Boulder Dam project "on the grounds that it involves Federal ownership and the sale of power." See ibid., p. 209.
50. J.C. Dawes, "To Improve Public Cleansing, Recognize It as a Science," *Municipal Sanitation* 2, no. 10 (Oct. 1931), p. 495. See also Soper, "The First International Conference"; "Some Financial, Political and Sanitary Phases of Garbage Disposal," *Engineering News-Record* 45, no. 7 (Feb. 14, 1901).
51. Desmond P. Tynan, "Modern Garbage Disposal—Incineration or Burial?: A Critical Study of the Sanitary Aspects and Costs of Refuse Disposal," *American City* 54, no. 6 (June 1939), p. 100; Sicular, "Currents in the Waste Stream," p. 15.
52. Caroline L. Hunt, "Home Problems from a New Standpoint: More Pleasure for the Producer of Household Stuff," *The Chautauquan* 37, no. 2 (May 1903), p. 179.
53. Laporte, *History of Shit*, pp. 46–47.
54. See Dolores Hayden, *The Grand Domestic Revolution: A History of Feminist Designs for American Homes, Neighborhoods, and Cities* (Cambridge, Mass.: MIT Press, 1981), pp. 243, 346–55.
55. Ibid., p. 205.
56. Charlotte Perkins Gilman, *The Home: Its Work and Influence* (New York: Charlton, 1910), p. 121.
57. Ibid., pp. 118–19.
58. "One Kitchen Fire for 200 People: No Necessity Any More for Each Family to Cook Its Own Meals," *Ladies' Home Journal* 35 (Sept. 1918), p. 97

59. Zona Gale, "Shall the Kitchen in Our Home Go?" *Ladies' Home Journal* 36 (March 1919), pp. 35, 50.

60. Hayden, *Grand Domestic Revolution*, pp. 281–83.

61. Ibid., pp. 275, 285–89.

62. Sicular, "Currents in the Waste Stream," p. 74. Also see "Street Cleaning Standards," *Municipal Journal and Public Works* 35, no. 24 (Dec. 11, 1913), p. 794.

Chapter 4

1. Louis Blumberg and Robert Gottlieb, *War on Waste: Can America Win Its Battle with Garbage?* (Washington, D.C.: Island Press, 1989), p. 6; Melosi, *Garbage in the Cities*, pp. 151, 164.

2. Some officials used land disposal in the short term while they saved up funds to buy capital-intensive treatment equipment, like incinerators. See E.J. Cleary, "Land Fills for Refuse Disposal," *Engineering News-Record* 121, no. 9 (Sept. 1, 1938), p. 273.

3. Blumberg and Gottlieb, *War on Waste*, p. 8. Incineration plants were under construction and renovation across the country in large and small municipalities alike, from Peekskill, New York, to New Orleans and Minneapolis, and even in "the model suburban village" of Greendale, Wisconsin. See "Technical Aspects of Refuse Disposal," *Civil Engineering* 9, no. 3 (March 1939), pp. 169–70.

4. "The Round Table," *Municipal Sanitation* 10, no. 4 (April 1939), p. 245.

5. Henry W. Taylor, "Power and Heat Production Feature in Incineration," *Municipal Sanitation* 9, no. 1 (Jan. 1938), p. 72. Neighborhood protests successfully halted the operation of an incinerator in San Francisco in 1929 through a court ruling that deemed the plant a nuisance. See John J. Casey, "Disposal of Mixed Refuse by Sanitary Fill Method at San Francisco," *Civil Engineering* 9, no. 10 (Oct. 1939), p. 590.

6. Taylor, "Power and Heat Production Feature in Incineration," p. 72 (italics added).

7. Melosi, *Garbage in the Cities*, p. 187. New Deal agencies like the Public Works Administration made immense investments in power plants. According to one report, sixty-one power plant construction projects in twenty-three states costing $117 million were almost fully financed with PWA grants and loans. In this climate alternative sources of power were not a high priority. See "ENR News of the Week: End of Litigation Sought by PWA," *Engineering News-Record* 120, no. 2 (Jan. 13, 1938), p. 43.

8. "An important element in incinerator cost is labor, at the incinerator more labor is required and a higher rate is paid than on the fill." See "Sanitary Fill Disposal Method Preferred at Portland," *Engineering News-Record* 123, no. 13 (Sept. 28, 1939),

p. 58. Also see "Salvaging Municipal Refuse in Three Cities," p. 24; "The Round Table," pp. 245–47.

9. "Sanitary Fill Disposal Method Preferred at Portland," p. 58. For each ton sent into the flames Erie, Pennsylvania paid dearly at $4.69, while Milwaukee forked over $2.37 and Detroit spent $1.40. See "The Round Table," pp. 245–47.

10. George M. Wisner, chief engineer, and Langdon Pearse, division engineer, *Report on the Pollution of Des Plaines River and Remedies Therefor* (Chicago: Press of Barnard and Miller, 1914), p. 6.

11. The first federal ban on ocean dumping, the Marine Protection Act, was passed in 1888. See Benjamin Miller, *Fat of the Land: Garbage of New York—The Last Two Hundred Years* (New York: Four Walls Eight Windows, 2000), p. 71; on the 1933 ban see Blumberg and Gottlieb, *War on Waste*, p. 7; Committee on Refuse Disposal, American Public Works Association, "Sanitary Landfills," *Municipal Refuse Disposal* (Chicago: Public Administration Service, 1961), p. 86.

12. Hering and Greeley, *Municipal Refuse*, p. 241.

13. On reduction plant economic strains and closures, see "Technical Aspects of Refuse Disposal," *Civil Engineering*, p. 170; Taylor, "Power and Heat Production Feature in Incineration," p. 72. On the effects of the falling price of grease, see "Salvaging Municipal Refuse in Three Cities," p. 23; Joseph E. Gill, "Garbage Reduction and Incineration Combined in Philadelphia Plant," *American City* 50, no. 12 (Dec. 1935), p. 58. On Philadelphia's plant closing, see Melosi, *Garbage in the Cities*, p. 217.

14. Roger J. Bounds, "A Survey of Practices in Refuse Disposal in American Cities," *Municipal Sanitation* 2, no. 9 (Sept., 1931), p. 433.

15. Grand Rapids, St. Paul, Omaha and Denver were some of the other municipalities employing the hog feeding method of garbage disposal during the 1930s. See "The Round Table," pp. 245–47; "Technical Aspects of Refuse Disposal," p. 166.

16. At the same time, 83 percent of Pacific Coast cities disposed of their organic offal by hog feeding. See Sicular, "Currents in the Waste Stream," p. 89; Melosi, *Garbage in the Cities*, p. 170; "Hog Feeding Dominates in Garbage Disposal," *Engineering News-Record* 123, no. 11 (Sept. 14, 1939), p. 72.

17. Bounds, "Survey of Practices in Refuse Disposal," p. 434. On Massachusetts's sixty-one towns and cities that used hog feeding for garbage disposal, see Melosi, *Garbage in the Cities*, p. 170.

18. All forty-four cities in Los Angeles County disposed of their organic waste by selling it to piggeries in 1930. See C.G. Gillespie and E.A. Reinke, "Municipal Refuse Problems and Procedures," *Civil Engineering* 4, no. 9 (Sept. 1934), pp. 487–91; "Salvaging Municipal Refuse in Three Cities."

19. Melosi, *Garbage in the Cities*, pp. 215–16; Sicular, "Currents in the Waste Stream," pp. 91–92; Willard H. Wright, "The Whole Truth about Hog Feeding," *Municipal Sanitation* 10, no. 5 (May 1939), pp. 268–70.
20. "Sanitation Progress Review: Lansing Pioneers in Joint Digestion of Sludge and Garbage," *Municipal Sanitation* 10, no. 1 (Jan. 1939), pp. 37–38.
21. C.E. Keefer, "Sewage System Utilized for Disposal of Garbage," *Engineering News-Record*. 112, no. 7 (Feb. 15, 1934), p. 227.
22. Ibid.; Bounds, "Survey of Practices in Refuse Disposal," p. 432.
23. The other municipalities that built composting systems were Dunedin, New York, and Belleair and Plant City, Florida. In 1932 the Netherlands operated a full-scale composting facility and since the 1960s about 2,500 municipal composting plants have opened in India, with another 100 in countries outside the United States. See Melosi, *Garbage in the Cities*, p. 221. Also see Sicular, "Currents in the Waste Stream," p. 95.
24. "Dumping City Refuse," *Municipal Journal and Engineer* 42 (Jan. 25, 1917), p. 103.
25. "From the 1880s to the 1930s, land dumping remained the most prevalent method of waste disposal." See Blumberg and Gottlieb, *War on Waste*, p. 7.
26. "The Land Disposal of Garbage: An Opportunity for Engineers and Contractors," *Engineering News* 53, no. 14 (April 6, 1905), p. 368. Also see Hering and Greeley, *Municipal Refuse*, pp. 253–56.
27. Mixed burial consisted of "spreading layers of garbage about 12 inches in thickness, and covering over with layers of 18 to 24 inches of ashes, street sweepings, rubbish and earth." See Bounds, "A Survey of Practices in Refuse Disposal," p. 431.
28. Harrison P. Eddy, Jr., "Cautions Regarding Land-Fill Disposal," *Engineering News-Record* 121, no. 24 (Dec. 15, 1938), p. 766. On the influence of controlled tipping, see Melosi, *Garbage in the Cities*, p. 219; Blumberg and Gottlieb, *War on Waste*, p. 16.
29. Born in 1894 and raised in St. Louis, Vincenz earned his degree in civil engineering from Stanford. Having lived in a boxcar while working in San Francisco as a rodman at Southern Pacific, Vincenz went on to open his own practice and held membership in the country's most powerful professional organizations, the American Society of Civil Engineers among them, and served as president to both the American Public Works Association and the California League of Municipalities. See Sicular, "Currents in the Waste Stream," pp. 99–102.
30. "Garbage Disposal at Fresno Placed on Efficient Basis," *Engineering News-Record* 114, no. 17 (April 25, 1935), p. 593.

31. "Some Financial, Political and Sanitary Phases of Garbage Disposal," *Engineering News-Record*, p. 121. By the 1930s rats were widely regarded as the source of much ill health. "Rats have been the indirect cause of the death of more human beings than have been caused by all the combined wars of history," wrote New York City's Department of Sanitation Commissioner William F. Carey rather hyperbolically. See Carey, "Comment and Discussion: Land Fills: Pro and Con," *Engineering News-Record* 121, no. 11, p. 317.
32. The switch to automated carts began in the teens, expanding throughout the succeeding decades. In 1912 Atlanta was experimenting with "gasoline driven trucks" that "proved very satisfactory." Around the same time Columbus, Boston, and Seattle were also getting motorized. See John H. Gregory, "Collection of Municipal Refuse," *American Journal of Public Health* 2, no. 12 (Dec. 1912), p. 921. Some municipalities stuck to a mixed fleet, relying on both new technology and tried and true collection methods. By the end of the 1930s Chicago employed horse-drawn trailers, tractors, and dump trucks; and Memphis put a combined force of one-ton trucks and single-mule carts to the task. See "Refuse Collection in 28 Cities in the United States," *American City* 53, no. 7 (July 1938), p. 59. Also see "The Round Table," *Municipal Sanitation* 2, no. 1 (Jan. 1931).
33. "Garbage Disposal at Fresno Placed on Efficient Basis," p. 593. Many other cities recognized the cost savings that could come from Taylorizing their refuse collection and disposal teams. See Cleary, "Land Fills for Refuse Disposal," p. 270.
34. "Garbage Disposal at Fresno Placed on Efficient Basis," pp. 592–93.
35. "Fresno's Garbage Plan Succeeds," *Engineering News-Record* 120, no. 10 (March 10, 1938), p. 365.
36. "With this plan of operation, only one man is required at the fill—i.e., the dragline operator, whose salary is $160 per month, constitutes the only labor charge at the fill." See "Garbage Disposal at Fresno Placed on Efficient Basis," p. 593.
37. Cleary, "Land-Fills for Refuse Disposal," p. 270. For more on types of land used for fills, see Leo King Couch, "Proper Garbage Disposal an Effective Aid in Rat Control," *Municipal Sanitation* 2, no. 6 (June 1931). Engineers were anxious to combat mosquitoes in large part because they transmitted diseases like malaria ("a destroyer of mankind and of nations"). Previous efforts to control mosquitoes were labor-intensive, scattershot, and intensely shortsighted by today's standards. Consider New York's methods involving ditch digging in its swampy areas: "When shallow water is run into the trenches it is much easier for the department to apply crude oil, which prevents larvae from rising to the surface for air." The city had 600 miles of such trenches. See J.A. Le Prince, "Building Malaria out of a Community

Through Engineering Work Designed to Prevent Mosquito Propagation," *Municipal Sanitation* 2, no. 1 (Jan. 1931), pp. 10–13. Also see V.M. Ehlers, "Sanitation Scores in Dallas Levee Improvements," *Municipal Sanitation* 2, no. 2 (Feb. 1931), p. 77.
38. E.J. Cleary, "Land-Fills for Refuse Disposal," *Engineering News-Record* 121, no. 9 (Sept. 1, 1938), p. 273.
39. "A Five-Year Plan on Sanitary Fills," *Engineering News-Record* 123, no. 5 (Aug. 3, 1939), p. 65.
40. George S. Smith, "Systems of Collection and Disposal of Garbage in City of New Orleans," *American Journal of Public Health* 2, no. 12 (Dec. 1912), p. 924.
41. Gordon M. Fair, "Comment and Discussion: Land Fill: Pro and Con," *Engineering News-Record* 121, no. 12 (Sept. 22, 1938), p. 347.
42. W.W. Harmon, quoted in Desmond P. Tynan, "Modern Garbage Disposal—Incineration or Burial?" *American City* 54, no. 6 (June 1939), p. 111.
43. "Technical Aspects of Refuse Disposal," p. 166.
44. "Fill Disposal of Refuse Successful in San Francisco," *Engineering News-Record* 123, no. 1 (July 6, 1939), p. 60; Casey, "Disposal of Mixed Refuse," pp. 590–91.
45. "San Francisco Garbage Disposal Continues With Fill-and-Cover on Tide-Flat Areas," *Engineering News-Record* 113, no. 16 (Oct. 18, 1934), p. 501.
46. Casey, "Disposal of Mixed Refuse," pp. 590–91.
47. Ibid., p. 591.
48. Eddy, "Cautions Regarding Land-Fill Disposal," p. 766.
49. Miller, *Fat of the Land*, p. 188. According to Sicular, Rikers Island started out at 60 acres and expanded with landfilling to 420 acres by the end of the 1930s. See Sicular, "Currents in the Waste Stream," p. 108.
50. Tynan, "Modern Garbage Disposal—Incineration or Burial?" p. 100. Also see Edward T. Russell, "Comment and Discussion: Land Fills: Pro and Con," *Engineering News-Record* 121, no. 13 (Sept. 29, 1938), p. 391.
51. *Staten Island Advance*, cited in Miller, *Fat of the* Land, p. 340.
52. On Moses's use of garbage for land reclamation, see ibid., pp. 191, 196. On protests, see Robert Moses, "Comment and Discussion: Land Fills: Pro and Con," *Engineering News-Record* 121, no. 11 (Sept. 15, 1938), p. 317; Miller, *Fat of the* Land, pp. 192–93.
53. Miller, *Fat of the Land*, pp. 188–98, 204. On Floyd Bennett Field, see www.geocities.com/floyd_bennett_field/1930s.html. On the World's Fair site, see Carlton S. Proctor, "Preparation of the Fair Site," *Engineering News-Record* 121, no. 12 (Sept. 22, 1938), pp. 353–56. La Guardia, built for the city by future sanitation commissioner Bill Carey (not Robert Moses) in 1938, was also constructed on land filled

in with waste. Causing problems beyond persistent rancid odor, the use of trash as a foundation for heavy construction in this case turned into a fiasco. The fill was so unstable and infested with rats that by the late 1940s the main terminal had "all but collapsed" and its runways, home to huge underground rat populations, were constantly pocked with sinkholes, requiring regular and intensive maintenance. See Miller, *Fat of the Land,* p. 187.

54. W. Earle Andrews, "New York World's Fair 1939: Its Background and Objective," *Engineering News-Record* 121, no. 12 (Sept. 22, 1938), p. 350.

55. Walter Benjamin, "Paris, Capital of the Nineteenth Century," in *Reflections*, edited by Peter Demetz (New York: Harcourt Brace, 1978), p.151.

56. Sicular, "Currents in the Waste Stream," pp. 115–16.

57. Ibid., p. 117.

58. Committee on Refuse Disposal, American Public Works Association, "Sanitary Landfills," p. 87.

59. Sicular, "Currents in the Waste Stream," p. 117.

60. Ibid., p. 118.

61. Ibid., p. 119.

62. Committee on Refuse Disposal, American Public Works Association, "Sanitary Landfills," p. 87.

63. Blumberg and Gottlieb, *War on Waste*, p. 17.

64. Richardson Wright, "The Decay of Tinkers Recalls Olden Days of Repairing," *House & Garden* 58 (Aug. 1930), p. 48.

65. Jean Vincenz made this statement at the 1940 American Public Works Association Congress. Quoted in Sicular, "Currents in the Waste Stream," p. 105.

66. Jean Vincenz, quoted in ibid.

67. Ibid., p. 104.

68. So benign in fact that an article in *Municipal Sanitation* declared: "Fills of this nature can often be made in exclusive residential districts." See Couch, "Proper Garbage Disposal an Effective Aid in Rat Control," p. 280. Other cities also grasped the benefits of masking the mess. In Seattle, one observer wrote, "An important feature of fill management in residential districts is the use of a screen made with either a row of quick-growing trees, such as willow or cottonwood, or with lattice fences painted green. Screened from direct view and with the health department supervising sanitary conditions, these fills are being successfully operated in all parts of the city." See "A Five-Year Plan on Sanitary Fills," *Engineering News* 123, no. 5 (Aug. 3, 1939), p. 66.

69. "Five Hundred Serve a Million and a Quarter," *American City* 50, no. 3 (March 1935), p. 47.

70. Ibid., p. 48.
71. "No attempt is made to salvage articles that have a value as junk, except by the collectors themselves." See Casey, "Disposal of Mixed Refuse," p. 590. Also see "Garbage Collection by Contract—Costs," *American City* 50, no. 10 (Oct. 1935), p. 15.
72. Eddy, "Cautions Regarding Land-Fill Disposal," p. 767. Also see "Technical Aspects of Refuse Disposal," p. 166.
73. "Technical Aspects of Refuse Disposal," p. 167.
74. For example, in San Francisco total daily collections of mixed refuse shrank from 1932 levels of 650–700 tons to 550 tons just two years later. See "San Francisco Garbage Disposal," p. 501. Another observation from the time: "It is interesting to observe on a quantitative basis the gradual reduction, during the current depression, in the food value of garbage. The depression has affected the purchasing power and therefore the economical instincts of the housewife in the matter of discarding food." See Herman Courtelyou, "Our Readers Say—Garbage Disposal in Los Angeles," *Civil Engineering* 5, no. 1 (Jan. 1935), p. 30. Also see Gillespie and Reinke, "Municipal Refuse Problems and Procedures."
75. American Public Works Association, quoted in Sicular, "Currents in the Waste Stream," p. 94.
76. Cleary, "Land-Fills for Refuse Disposal," p. 273; "A Five-Year Plan on Sanitary Fills," *Engineering News* , p. 66; Eddy, "Cautions Regarding Land-Fill Disposal," p. 766.
77. Strasser, *Waste and Want*, pp. 140–41.
78. Ibid., p. 141.
79. Ibid., p. 262.

Chapter 5

1. U.S. EPA, Office of Solid Waste and Emergency Response, *Municipal Solid Waste in the United States: 2001 Facts and Figures Executive Summary* (Washington, D.C., Oct. 2003), p. 5.
2. Concentration continued into the post–World War II years: "In 1950, out of some 300,000 manufacturing firms in the United States, the first five companies produced some 12 per cent of the total value of products manufactured." See Victor Lebow, "Forced Consumption—The Prescription for 1956," *Journal of Retailing* 31, no. 4 (Winter 1955/56), p. 168.
3. The laws of competition meant that "consumer goods manufacturers were coming to recognize that mass production and mass distribution were 'necessary'

steps toward survival in a competitive market." See Stuart Ewen, *Captains of Consciousness: Advertising and the Social Roots of the Consumer Culture* (New York: Basic Books, 2001 [1976]), p. 24.

4. Harold C. Livesay, *Andrew Carnegie and the Rise of Big Business* (New York: Longman, 2000), p. 130.

5. The NAWMD tasked itself with classifying scrap materials and connecting waste buyers and sellers, as well as giving background credit information and providing free arbitration for any resulting trade disputes. See National Association of Waste Material Dealers, *Twenty-Fifth Anniversary Blue Book, 1913–1938* (New York: NAWMD, 1938), pp. 13–21; National Association of Waste Material Dealers, *Fifteenth Anniversary Blue Book, 1913–1928* (New York: NAWMD, 1928), p. 49. A plethora of other large-scale, membership-based salvage and refuse brokers set up shop in the first three decades of the twentieth century. See Strasser, *Waste and Want*, p. 118.

6. "The WPB had authority to convert factories, set quotas for war production, determine the supply of materials to industry, and issue orders for conserving materials and limiting industrial production." See Strasser, *Waste and Want*, p. 238.

7. Ibid., pp. 238–39.

8. David Harvey, *The Condition of Postmodernity* (Oxford: Blackwell, 1990), p. 127.

9. Ibid., p. 129.

10. The goal of scientific management was to reduce inefficiencies by streamlining labor processes. This rationalized system got raw materials in and the end product out as quickly and cost-effectively as possible, thereby rendering workers super-efficient.

11. Harvey, *Condition of Postmodernity*, p. 133.

12. Quoted in Strasser, *Waste and Want*, pp. 231–32. As a measure of wartime consumer restraint, from 1941 to 1945 personal savings reached an average of 21 percent of disposable income, compared to just 3 percent in the 1920s. See Lizabeth Cohen, *A Consumers' Republic: The Politics of Mass Consumption in Postwar America* (New York: Knopf, 2003), pp. 70–71.

13. "[T]he liquid assets of individuals nearly tripled between Pearl Harbor and the end of the war." See Strasser, *Waste and Want*, p. 233. For more on wartime income and savings, see Cohen, *Consumers' Republic*, pp. 69–71.

14. Elaine Tyler May, "The Commodity Gap," in *Consumer Society in American History: A Reader*, edited by Lawrence B. Glickman (Ithaca, N.Y.: Cornell University Press, 1999), p. 301.

15. Ibid.

16. "As a result of all these inducements, housing starts went from 114,000 in 1944 to an all-time high of 1,692,000 in 1950." See May, "Commodity Gap," pp. 303–4. These programs had roots in previous interventions on behalf of business, like Hoover's Conference on Home Building and Home Ownership of 1931. Backed by developers and manufacturers, the program threw federal support behind a national home ownership scheme. See Hayden, *Grand Domestic Revolution*, p. 286.
17. As Nixon stated in the debate: "Thirty-one million families own their own homes and the land on which they are built. America's 44 million families own a total of 56 million cars, 50 million television sets and 143 million radio sets." See May, "Commodity Gap," p. 299.
18. In 1930, just 24 percent of U.S. households owned washing machines, but by 1960 that number had shot up to 73 percent. And the proportion of U.S. families with cars increased from 58 percent in 1942 to 75 percent in 1960. See Stanley Lebergott, *Pursuing Happiness: American Consumers in the Twentieth Century* (Princeton, N.J.: Princeton University Press, 1993), pp. 115, 130.
19. Paul Mazur, *The Standards We Raise: The Dynamics of Consumption* (New York: Harper & Brothers, 1953), pp. 19–20.
20. Vance Packard, *The Waste Makers* (New York: Pocket Books, 1960), p. 27. Some features added to commodities were purely a sham to lure in customers, "like the first refrigerator cold controls, which were not connected to anything." See Martin Mayer, "Planned Obsolescence: Rx for Tired Markets?" *Dun's Review and Modern Industry* 73, no. 2 (Feb. 1959), p. 80. Also, some refrigerator and radio manufacturers experimented with regular stylistic and technological changes in the interwar years, but not to the degree achieved after the war. See Strasser, *Waste and Want*, p. 195.
21. All quoted in Packard, *Waste Makers*, pp. 25–27.
22. Packard, *Waste Makers*, p. 25.
23. Karl Marx, *Grundrisse: Foundations of the Critique of Political Economy*, translated by Martin Nicolaus (New York: Penguin Books, 1993), p. 408.
24. Quoted in Packard, *Waste Makers*, p. 21.
25. Well before the 1950s, manufacturers who simply built products to last struggled against competitive firms who innovated new and changing styles to attract consumer attention and dollars. Henry Ford—who manufactured the Model T "so strong and so well-made that no one ought ever to have to buy a second one"—was forced to introduce varying styles and colors after GM began marketing changing designs in the mid-1920s. See Strasser, *Waste and Want*, p. 194.
26. Christine Frederick, *Selling Mrs. Consumer* (New York: Business Bourse, 1929), p. 246.

27. Internal documents revealed in 1939 that GE officials instructed manufacturers to reduce lamp life. See Packard, *Waste Makers*, p. 51.
28. Quoted in Packard, *Waste Makers*, p. 56.
29. Ibid.
30. Between 1957 and 1958, although there were only 1 million new cars on the road, the AAA reported a record 5.5 million breakdowns, most likely due to shoddy parts. Ignition manufacturer quoted in Packard, *Waste Makers*, pp. 80–81.
31. Quoted in Packard, *Waste Makers*, p. 10.
32. Ibid., pp. 69, 75.
33. Victor Lebow, "Price Competition in 1955," *Journal of Retailing* 31, no. 1 (Spring 1955), p. 7.
34. From the documentary film *Gone Tomorrow: The Hidden Life of Garbage*, directed by Heather Rogers, 2002. The term "creative destruction" refers to Joseph A. Schumpeter's analysis that capitalism is constantly changing from within as new systems destroy the old ones; for example, new industrial processes are incessantly replacing established ones. See Schumpeter, *Capitalism, Socialism and Democracy* (New York: Harper & Row, 1976 [1942]), pp. 81–86.
35. E.S. Safford, "Product Death-Dates—A Desirable Concept?" *Design News* 13, no. 24 (Nov. 24, 1958), p. 3.
36. While disposable diapers have made life easier for women, they are not the only solution. If industry invested as much on researching and improving reusable diapers as they have on disposables, there might be more realistic options today. See Louis Blumberg and Robert Gottlieb, *War on Waste: Can America Win Its Battle with Garbage?* (Washington, D.C.: Island Press, 1989), p. 247. On the Dryper TV advertisement, see the documentary film *Gone Tomorrow*.
37. On the first disposable blades, see Susan Strasser, *Satisfaction Guaranteed: The Making of the American Mass Market* (New York: Ballentine, 1989), p. 101. In 1976 Gillette introduced its first disposable twin-blade razor for men, which it called "Good News!" With this new version, shavers pitched the whole commodity, not just the dull blade. See www.gillette.com/men/features/100years.htm.
38. Packard, *Waste Makers*, pp. 37–38.
39. Walter Stern quoted in Blumberg and Gottlieb, *War on Waste*, p. 12. For more on strip malls and chain stores, see Gordon H. Stedman, "The Rise of Shopping Centers," *Journal of Retailing* 31, no. 1 (Spring 1955).
40. Blumberg and Gottlieb, *War on Waste*, p. 12. Before the advertising boom, "packaging was functional, not a strategic element in a successful sales campaign." Ibid., p. 11.

41. On tobacco tins, see Strasser, *Waste and Want*, p. 172.
42. "Corporations: Growing Package," *Time* 73 (Jan. 5, 1959), pp. 76–78.
43. Packard, *Waste Makers*, p. 39.
44. Roy Sheldon and Egmont Arens, *Consumer Engineering: A New Technique for Prosperity* (New York: Harper Brothers, 1932), p. 55.
45. The average annual growth rate measured between 1948 and 1970. See Jeffrey L. Meikle, *American Plastic: A Cultural History* (New Brunswick, N.J.: Rutgers University Press, 1997), p. 265.
46. Alan Hess, "Monsanto House of the Future," *Fine Homebuilding* 34 (Aug. / Sept. 1986), p. 75.
47. Meikle, *American Plastic*, p. 5.
48. Using the injection molding process. See ibid., p. 29.
49. Roland Barthes, *Mythologies* (New York: Noonday Press, 1990 [1957]), pp. 98–99.
50. Meikle, *American Plastic*, p. 103.
51. Stephen Fenichell, *Plastic: The Making of a Synthetic Century* (New York: HarperCollins / Harper Business, 1996), p. 203.
52. A few decades after the war a writer discovered Bakelite's use in atomic weaponry, but was restricted by the U.S. Atomic Energy Commission from releasing his research. "Whatever arcane contribution Bakelite made to the development of the atom bomb remains classified to the present day." See Fenichell, *Plastic*, p. 102. On other wartime uses, see Meikle, *American Plastic*, p. 1.
53. Before Saran film came along, the military had to disassemble equipment, pack it in grease for protection from water during shipping, then degrease and reassemble upon arrival. See Fenichell, *Plastic*, pp. 211–12.
54. Ibid., p. 193.
55. Meikle, *American Plastic*, p. 160.
56. Joseph L. Nicholson and George R. Leighton, "Plastics Come of Age," *Harper's Magazine* 185, no. 1107 (Aug. 1942), pp. 300–301.
57. Felix N. Williams, "The Peak Is Not Yet," *Monsanto Magazine* 26, no. 4 (Oct. 1947), p. 2. Injection molding machines were pioneered in Germany and first introduced to American production lines in the 1920s as a more efficient fabricator for the malleable heat softening "cellulose acetate." Injection molding of plastic was far more productive than the previously dominant method called compression molding and was continually upgraded during and after World War II. See F.A. Abbiati, "The Horn of Plenty Is Mechanized," *Monsanto Magazine* 26, no. 4 (Oct. 1947), p. 27.

58. The author goes on: "In the thirty seconds it took you to read this far, a modern injection molding machine turned out sixteen combs from a single mold in one operation" (Abbiati, "Horn of Plenty," p. 26). Also see Hiram McCann, "Doubling—Tripling—Expanding: That's Plastics," *Monsanto Magazine* 26, no. 4 (Oct. 1947), p. 4.
59. Meikle, *American Plastic*, p. 1.
60. James W. Sullivan, general manager of Union Products, quoted in Meikle, *American Plastic*, p. 180.
61. "Polyethylene Meets New Markets with King Size Moldings," *Modern Plastics* 34, no. 2 (Oct. 1956), p. 123; "Loma's Barnett: A Flair for 'Firsts,'" *Modern Plastics* 39, no. 9 (May 1962), p. 48.
62. Lloyd Stouffer, editor of *Modern Packaging*, quoted in "Plastics in Disposables and Expendables," *Modern Plastics* 34, no. 8 (April 1957), p. 93; emphasis in original. *Modern Packaging* magazine was owned by the publisher of *Modern Plastics*.
63. "Plastics for Disposables," *Modern Plastics* 33, no. 8 (April 1956), p. 5; emphasis in original.
64. "Plastics in Disposables and Expendables," p. 96.
65. Meikle, *American Plastic*, p. 187.
66. Ibid., p. 189. On the wartime and postwar uses of Styrofoam (a trademark of Dow Chemical), see O.R. McIntire, "Styrofoam," in *A History of the Dow Chemical Physics Lab: The Freedom to Be Creative*, ed. Ray H. Boundy and J. Lawrence Amos (New York: Marcel Dekker, Inc., 1990), pp. 117–28.
67. "Plastics in Disposables and Expendables," pp. 93–99.
68. William F. Cullom, "Wrapped in Plastics Films," *Modern Plastics* 25, no. 9 (May 1948), p. 87.
69. Advertisement, *Modern Plastics* 33, no. 8 (April 1956), p. 283.
70. Meikle, *American Plastic*, pp. 265–66.
71. "Plastics in Disposables and Expendables," p. 96.
72. A former *Modern Plastics* editor, Sidney Gross, observed that if manufacturers use plastics, consumers have no choice but to use them. See Meikle, *American Plastic*, p. 275.
73. Ibid., p. 266.
74. *Printers' Ink*, quoted in Ewen, *Captains of Consciousness*, p. 53; italics added.
75. Blumberg and Gottlieb, *War on Waste*, p. 15.
76. On 1920 figure, see Hayden, *Grand Domestic Revolution*, p. 274. On 1950 and 1956 figures, see William W. Keep, Stanley C. Hollander and Roger Dickinson, "Forces Impinging on Long-Term Business-to-Business Relationships in the United

States," *Journal of Marketing* 62, no. 2 (April 1, 1998).
77. Baxandall and Ewen, *Picture Windows*, p. 134.
78. Victor Lebow, "Forced Consumption—The Prescription for 1956," p. 169.
79. Keep, Hollander, and Dickinson, "Forces Impinging on Long-Term Business-to-Business Relationships."
80. On early marketing see Strasser, *Satisfaction Guaranteed*, pp. 5–7.
81. "Under private property . . . every person speculates on creating a *new* need in another, so as to drive him to a fresh sacrifice, to place him in a new dependence and to seduce him into a new mode of *gratification* and therefore economic ruin." See Karl Marx, "The Economic and Philosophic Manuscripts of 1844," in *The Marx-Engels Reader*, edited by Robert C. Tucker (New York: W.W. Norton, 1978), p. 93; emphasis in original.
82. This midcentury trend of separating the consumer from class consciousness began earlier in the century. See Stuart Ewen and Elizabeth Ewen, *Channels of Desire: Mass Images and the Shaping of American Culture* (Minneapolis: University of Minnesota Press, 1992), p. 265.
83. John Berger, *Ways of Seeing* (Harmondsworth: Penguin, 1972), pp. 148–49. Stuart Ewen has made a similar point: "If man was the victim of himself, the fruits of mass production were his savior." See Ewen, *Captains of Consciousness*, p. 46.
84. Ewen and Ewen, *Channels of Desire*, p. 262.
85. On the "Smokatron," see Committee on Refuse Disposal, American Public Works Association, *Municipal Refuse Disposal* (Danville, Ill.: Interstate, 1961), p. 80.
86. On the "In-Sink-Erator," see www.insinkerator.com/history.html.
87. New York City did not legalize in-sink grinders until 1997. See Jay Romano, "Your Home; Grinding Garbage in the Sink," *New York Times*, Dec. 7, 1997; Kendall Christiansen, "Food Disposers Help Grind Down Solid Waste Problems," *World Wastes* 41, no. 11 (Nov. 1, 1998); "Tax Rebates Approved for Installation of Kitchen Food Waste Disposers," *Real Estate Weekly* 47, no. 33 (March 21, 2001).
88. Committee on Refuse Disposal, American Public Works Association, *Municipal Refuse Disposal*, pp. 56–57.
89. Ibid., p. 55.
90. Blumberg and Gottlieb, *War on Waste*, p. 199.
91. Melosi, *Garbage in the Cities*, p. 208.

Chapter 6

1. Gladwin Hill, "Earth Day Goals Backed by Hickel," *New York Times*, April 22, 1970.

2. All quoted in Barry Commoner, *The Closing Circle: Nature, Man and Technology* (New York: Knopf, 1971), pp. 7-9.

3. African American students criticized the action, suggesting their white middle-class colleagues should have donated the $2,500 to a scholarship fund instead. See Commoner, *Closing Circle*, p. 207.

4. Hill, "Hickel."

5. The New York incidents occurred in 1963 when 405 people died, and in 1966 when another 168 were killed. See Dr. Richard Wade, "Health Effects: Pick Your Poison," *Environmental Action* 6, no. 24 (April 26, 1975), p. 5. These types of incidents tapered off after air pollution laws were passed, in Britain in the mid-1950s and in the United States in the mid-1960s. Since 1970 emissions of soot particles have been cut by nearly 80 percent in the United States. See www.epa.gov/region2/epa30/air.htm.

6. John Stauber and Sheldon Rampton, *Toxic Sludge Is Good for You: Lies, Damn Lies and the Public Relations Industry* (Monroe, Me.: Common Courage Press, 1996), pp. 123–25. Tragically and ironically, Carson died in 1964 from cancer.

7. Peter Wild, *Pioneer Conservationists of Western America* (Missoula, Mont.: Mountain Press, 1979), p. 156.

8. Quoted in Barry Commoner, *Making Peace with the Planet* (New York: New Press, 1992 [1975]), p. 173.

9. Quoted in Commoner, *Making Peace*, p. 181. On creation of the EPA and the Resource Recovery Act, see Blumberg and Gottlieb, *War on Waste*, p. 63.

10. Center for Investigative Reporting and Bill Moyers, *Global Dumping Ground: The International Traffic in Hazardous Waste* (Washington, D.C.: Seven Locks Press, 1990), p. 9.

11. John Quarles, "Fighting the Corporate Lobby," *Environmental Action* 6, no. 15 (Dec. 7, 1974), p. 5.

12. By 1974 beverage container waste was growing at a rate of 8 percent annually. See Catherine Lerza, "Administration 'Pitches In' to Outlaw Throwaways," *Environmental Action* 6, no. 2 (May 25, 1974), p. 5.

13. Thomas W. Fenner and Randee J. Gorin, *Local Beverage Container Laws: A Legal and Tactical Analysis* (Stanford, Calif.: Stanford Environmental Law Society, 1976), p. 3.

14. Packaging comprised 34 percent of municipal discards. See ibid., p. 2. Packaging had grown massively over the preceding decade: in 1960 per capita daily consumption was 0.73 pounds; ten years later it had surged to 1.05 pounds. See Blumberg and Gottlieb, *War on Waste*, p. 13.

15. Some estimates set this figure as low as fifteen; others set it as high as forty.
16. Fenner and Gorin, *Local Beverage Container Laws*, p. 4.
17. "Wrap-Up for Glass," *Modern Packaging* 34, no. 8 (April 1961), p. 289.
18. Kenneth C. Fraundorf, "The Social Costs of Packaging Competition in the Beer and Soft Drink Industries," *Antitrust Bulletin* 20 (Winter 1975), pp. 810–16.
19. Peter Chokola quoted in Lerza, "Administration 'Pitches In,'" p. 5. Also see Commoner, *Making Peace with the Planet*, p. 106.
20. Lerza, "Administration 'Pitches In,'" p. 5.
21. "Wrap-up for Glass," pp. 156–59, 287, 289.
22. Fraundorf, "Social Costs of Packaging Competition," p. 813.
23. Abbie Hoffman, *Steal This Book* (New York: Four Walls Eight Windows, 1996 [1971]), pp. 127–33.
24. Ibid., p. 24.
25. Patricia Taylor, "Source Reduction: Stemming the Tide of Trash," *Environmental Action* 6, no. 7 (Aug. 17, 1974), pp. 8–11.
26. GrassRoots Recycling Network, *Wasting and Recycling in the United States 2000*, p. 30; Blumberg and Gottlieb, *War on Waste*, pp. 199–200.
27. From an interview with Neil Seldman of the Institute for Local Self Reliance, Jan. 20, 2005.
28. Crucially, public officials were not typically involved in establishing these early recycling efforts. See Blumberg and Gottlieb, *War on Waste*, p. 19.
29. Quoted in Taylor, "Source Reduction," p. 9.
30. Ibid., p. 11.
31. On resistance to built-in-obsolescence, see Peter Harnik, "The Junking of an Anti-Litter Lobby," *Business and Society Review* 21 (Spring 1977), p. 48. On Wyeth, see Meikle, *American Plastic*, p. 266. On the introduction of PET, see Blumberg and Gottlieb, *War on Waste*, p. 237.
32. John H. Fenton, "Vermont's Session Has Budget Clash," *New York Times*, Feb. 1, 1953.
33. Blumberg and Gottlieb, *War on Waste*, p. 238.
34. On "one-way" inventors, see Fraundorf, "The Social Costs of Packaging Competition," pp. 806n10, 807. NAM's PR department clearly influenced KAB's approach; see Edward Maher, "NAM's Approach to Public Relations," *Public Relations Journal* 17, no. 5 (May 1961), pp. 4–6. On Coca-Cola and KAB, see Rebecca McCarthy, "Recycling: Will It Survive?" *Atlanta Journal-Constitution*, Dec. 6, 1993.
35. Within its first year, KAB was already working with civic organizations in ten different states and had announced a "plan to enlist the aid of governors in estab-

lishing committees in their states to cooperate with the national program." See "Heads Anti-Litter Unit," *New York Times*, Nov. 20, 1954. On ten states, see "Industry Fosters Drive against Litter," *New York Times*, Oct. 14, 1954.
36. Bernard Stengren, "What Makes a Litterbug?" *New York Times*, Dec. 5, 1954.
37. "Progress Is Noted in U.S. Clean-up," *New York Times*, Feb. 24, 1959.
38. Charles Grutzner, "Crackdown on Litterbugs," *New York Times*, Oct. 16, 1955. Some states, like Georgia, Massachusetts, California, Indiana, Maryland, Delaware and Nebraska already had antilitter laws that fined individuals for discarding refuse in unsanctioned sites.
39. "Vermont's Bottle Law Dies," *New York Times*, April 5, 1957.
40. While litter had been a topic of discussion for Progressives and urban beautifiers since the nineteenth century, it did not have quite the political and psychological power that KAB instilled in the term.
41. *Heritage of Splendor*, Alfred Higgins Productions, produced by Richfield Oil Corporation for Keep America Beautiful, Inc., 1963. See www.archive.org/movies.
42. Quoted in Blumberg and Gottlieb, *War on Waste*, p. 19.
43. Stuart Ewen, *PR! A Social History of Spin* (New York: Basic Books, 1996), p. 376.
44. Raymond Geuss, *Public Goods, Private Goods* (Princeton, N.J.: Princeton University Press, 2001), p. 18.
45. The point of this argument is not that littering is acceptable (although one could argue that if trash was not so quickly and seamlessly hauled away it might raise consciousness about the high stakes of mass wasting). The real issue, rather, lies with the use of the device of litter to displace responsibility for environmental destruction away from industrial production and onto individual consumers.
46. "The Waste-High Crisis," *Modern Packaging* 41, no. 11 (Nov. 1968), p. 102.
47. Harnik, "The Junking of an Anti-Litter Lobby," p. 50. In addition, see Daniel Zwerdling, "Iron Eyes," *All Things Considered*, National Public Radio (Jan. 10, 1999).
48. Taylor, "Source Reduction," p. 11.
49. The slogan ran in a *San Francisco News* editorial on May 20, 1959, quoted in Meikle, *American Plastic*, p. 251.
50. Ibid., pp. 249–53.
51. Lerza, "Administration 'Pitches In,'" p. 3; Harnik, "The Junking of an Anti-Litter Lobby," p. 48.
52. Fenner and Gorin, *Local Beverage Container Laws*, pp. 11, 14.
53. Ibid., p. 13.

54. Frank Teagle, Jr., letter to the editor, *Environmental Action* 6, no. 17 (Jan. 18, 1974), p. 2.
55. Blumberg and Gottlieb, *War on Waste*, p. 239.
56. Quoted in Blumberg and Gottlieb, *War on Waste*, p. 19.
57. Ibid., pp. 276–77.
58. Quoted in Harnik, "The Junking of an Anti-Litter Lobby," p. 48.
59. Four of the random shoppers worked for major beverage distributors; one man was the operational director for Columbia Distributing, a longtime opponent of the Oregon law. The fifth interviewee worked for Safeway, another corporation firmly against mandatory deposits. See Blumberg and Gottlieb, *War on Waste*, pp. 226–28.
60. Fenner and Gorin, *Local Beverage Container Laws*, p. 13.
61. "That Makes Cents," *Environmental Action* 6, no. 20 (March 1, 1975), p. 13.
62. Harnik, "The Junking of an Anti-Litter Lobby," p. 49.
63. Ibid., p. 51.
64. Fenner and Gorin, *Local Beverage Container Laws*, p. 13. The third state to enact deposit legislation was South Dakota.
65. Blumberg and Gottlieb, *War on Waste*, p. 239.
66. "Comment," *Environmental Action* 6, no. 20 (March 1, 1975), p. 3.
67. Harnik, "The Junking of an Anti-Litter Lobby," p. 49.
68. Fenner and Gorin, *Local Beverage Container Laws*, p. 13.
69. Letter cited in "Comment," *Environmental Action*, p. 3.
70. See Karl Marx, "The Economic and Philosophic Manuscripts of 1844," in *The Marx-Engels Reader*, edited by Robert C. Tucker (New York: W.W. Norton, 1978), p. 72.
71. Roughly, M = money, C = commodity (within which lies the production of that item), and M'= profit. In very broad strokes, the equation illustrates the pathway of the creation of profits. The point about garbage being part of that formulation came out of multiple conversations with Christian Parenti and John Marshall.
72. From the documentary film *Gone Tomorrow*.
73. Quoted in Ewen, *PR! A Social History of Spin*, pp. 397–98.

Chapter 7

1. See Commoner, *Making Peace with the Planet*, p. 107.
2. This figure refers to prices between the years 1982 and 1987. See Cynthia Pollock, *World Watch Paper 76: Mining Urban Wastes: The Potential For Recycling* (Washington, D.C.: Worldwatch Institute, April 1987), p. 15.

3. On Superfund sites, see Seldman, "Recycling—History in the United States," p. 2354. On the Fresno landfill, see Danielle Jackson, "California Landfill's Landmark Status under Review," *Waste Age* 32, no. 10 (Oct. 1, 2001).

4. Blumberg and Gottlieb, *War on Waste,* pp. 64–65. According to waste utilization expert Neil Seldman, Subtitle D was implemented in the mid-1980s due to pressure and lobbying from the country's largest garbage corporations like Waste Management Inc. and Browning-Ferris Industries. It was in these firms' interest to have more costly controls on landfills because such regulations made building and operating a site too expensive for municipalities and smaller trash companies. From an interview, Jan. 20, 2005.

5. Center for Investigative Reporting and Moyers, *Global Dumping Ground*, p. 7.

6. Travis W. Halleman, *A Statistical Analysis of Wyoming Landfill Characteristics* (master's thesis, Department of Civil and Architectural Engineering, University of Wyoming, Aug. 2004). Another source reports that the number of landfills shrank from 1,500 in 1984 to a miniscule 325 just four years later. See Center for Investigative Reporting and Moyers, *Global Dumping Ground*, p. 7.

7. Seldman, "Recycling—History in the United States," p. 2352.

8. Seldman, "Recycling—History in the United States," p. 2356. According to Seldman, "The recycling movement was the direct outcome of the anti-incineration movement." From an interview, Jan. 20, 2005.

9. Daniel Imhoff, "Thinking Outside of the Box," *Whole Earth* (Winter 2002), p. 13.

10. Seldman, "Recycling—History in the United States," p. 2354. "Resource recovery" is also used to refer to a wide range of programs other than incineration like recycling, composting and reuse. As Barry Commoner has pointed out, monikers like waste-to-energy were—and still are—disingenuous and misleading since no curtailing or transformation of the wasting cycle actually occurrs. See Commoner, *Making Peace with the Planet*, p. 108.

11. Seldman, "Recycling—History in the United States," p. 2354.

12. Cited in Blumberg and Gottlieb, *War on Waste*, p. 39.

13. Ibid., p. 71.

14. Commoner, *Making Peace with the Planet*, p. 110. Although the first revelations that municipal waste incinerators released dioxins came in the late 1970s, it was not until the following decade that citizens began organizing against incinerators specifically due to dangerous dioxin emissions. See Lois Marie Gibbs and the Citizens Clearinghouse for Hazardous Waste, *Dying from Dioxin: A Citizens' Guide to Reclaiming Our Health and Rebuilding Democracy* (Boston: South End Press, 1995), p. *xxx*.

15. Commoner, *Making Peace with the Planet*, p. 115. On the formation of the Community Alliance for the Environment, see Miller, *Fat of the Land*, p. 277.
16. "It is now generally accepted, by the incinerator industry as well as government agencies, that dioxin is synthesized in trash-burning incinerators." See Commoner, *Making Peace with the Planet*, pp. 116–18. On the human health effects of dioxin, see Pollock, *Mining Urban Wastes*, pp. 17–18; Health Care Without Harm, "Dioxin Fact Sheet," www.greenpeace.org.au/toxics/pdfs/dioxin_facts.pdf. Also see Gibbs and the Citizens Clearinghouse for Hazardous Waste, *Dying from Dioxin*, p. 1. It wasn't until 1996 that the EPA formally confirmed dioxin's dangerous effects on human immune and reproductive systems. See GrassRoots Recycling Network, *Wasting and Recycling in the United States 2000*, p. 30.
17. Commoner, *Making Peace with the Planet*, pp. 118–19.
18. Ibid., pp. 110–15, 118.
19. Blumberg and Gottlieb, *War on Waste*, pp. 69-70.
20. Ibid., pp. 156, 168–70. The UCLA students, organized by Robert Gottlieb and Louis Blumberg, both professors at the school, conducted their own report on the plant's health risks, providing CCSCLA with hard evidence to argue their case. See Commoner, *Making Peace with the Planet*, p. 124.
21. Blumberg and Gottlieb, *War on Waste*, pp. 180–83.
22. Ibid., p. 183.
23. Commoner, *Making Peace with the Planet*, p. 125. On Philadelphia, see Seldman, "Recycling—History in the United States," p. 2359.
24. Quoted in Commoner, *Making Peace with the Planet*, p. 123.
25. Bill Richards, "Burning Issue: Energy from Garbage Loses Some of Promise," *Wall Street Journal*, June 16, 1988.
26. Commoner, *Making Peace with the Planet*, p. 139.
27. Jeff Chang and Lucia Hwang, "It's a Survival Issue: The Environmental Justice Movement Faces the New Century," *ColorLines* 3, no. 2 (Summer 2000).
28. Blumberg and Gottlieb, *War on Waste*, p. 208.
29. Ibid., p. 209.
30. Ibid., p. 210; Commoner, *Making Peace with the Planet*, pp. 131–33.
31. Strasser, *Waste and Want*, p. 285.
32. Although various companies make foamed polystyrene items—often generically referred to as "Styrofoam"—the name and the chemical mixture of Styrofoam are property of Dow Chemical. While Dow's trademarked brand of polystyrene is no longer used for packaging like McDonald's clamshells, egg cartons and disposable cups, it used to be. See McIntire, "Styrofoam," in *A History of the Dow Chemical*

Physics Lab, p. 128. On 99 percent of all plastics, see Blumberg and Gottlieb, *War on Waste*, p. 14. On the annual volume of plastic discarded, see Meikle, *American Plastic*, p. 267. Plastics took up 7 percent of all wastes by weight, but weight-based figures can be misleading when referring to plastic discards.

33. On McDonald's as the country's largest polystyrene consumer, see Blumberg and Gottlieb, *War on Waste*, p. 258. On the campaign, see the McSpotlight website, www.mcspotlight.org/campaigns/countries/usa/usa_toxics.html.

34. Blumberg and Gottlieb, *War on Waste*, pp. 258–59.

35. Ibid.; McSpotlight website.

36. Quoted in Lerza, "Administration 'Pitches In,'" p. 5.

37. The recycling symbol was designed by Gary Anderson in 1970 for the Container Corporation of America, a paper products manufacturer. See "The History of the Recycling Symbol," *Dyer Consequences!*, available from home.att.net/~DyerConsequences/recycling_symbol.html.

38. On NCRR's backers, see "Comment," *Environmental Action* 6, no. 20 (March 1, 1975); Judd H. Alexander, *In Defense of Garbage* (Westport, Conn.: Praeger, 1993), p. 115; Neil Seldman, "Recycling—History in the United States," p. 2354.

39. New Orleans signed the contract with NCRR in 1973. See Patricia Taylor, "Source Reduction: Stemming the Tide of Trash," *Environmental Action* 6, no. 7 (Aug. 17, 1974), p. 11.

40. Alexander, *In Defense of Garbage*, p. 115.

41. Ibid., p. 116.

42. Taylor, "Source Reduction," p. 11.

43. "Comment," *Environmental Action*, March 1, 1975.

44. On the newsletter, see Ecology Center, "Report of the Berkeley Plastics Task Force" (Berkeley, Calif., April 8, 1996), p. 7. On spending on packaging, see Pollock, *Mining Urban Wastes*, p. 8.

45. Still true today, plastic bottles are separated for recycling based not on their grade number but on their shape since the latter indicates how they were made. Once plastic is manufactured, it can only be recycled using that same production process, of which there are several. From an interview with Neil Seldman, Jan. 20, 2005.

46. The Ecology Center, *Report*, pp. 6–7.

47. Cited in ibid., p. 7.

48. GrassRoots Recycling Network, *Wasting and Recycling in the United States 2000*, p. 12.

49. Victor Wigotsky, quoted in Ecology Center, *Report*, p. 6.

50. On APC's opposition to legislation, see Ecology Center, *Report*, p. 6. Resulting from the APC's abandonment of its 25 percent recycling target, the group was given fines from eleven states attorneys general and forced to sign a consent decree as punishment for deceptive advertising. See GrassRoots Recycling Network, *Wasting and Recycling in the United States 2000*, p. 41.

51. Karl Marx, "Economic and Philosophic Manuscripts of 1844," in *The Marx-Engels Reader*, edited by Robert C. Tucker (New York: W.W. Norton, 1978), p. 103.

52. "U.S. Environmental Protection Agency figures indicate that the nation's municipal recycling level has stagnated at 28% in 1997, not much greater than the 27% level reported the previous year. And, for the first time since 1993, the tonnage landfilled and incinerated has increased both in absolute tons and on a per capita basis." See GrassRoots Recycling Network, *Wasting and Recycling in the United States 2000*, p. 10.

53. On more Americans recycling than voting, see ibid., p. 9.

54. Ibid., p. 13. Also see Jim Motavalli, "Zero Waste," *E/The Environmental Magazine* 12, no. 2 (March/April 2001), p. 28.

55. These facts come from several interviews with Dave Williamson of Ecology Center, conducted in Berkeley, California, in 2001–2002.

56. Ecology Center, *Report*, p. 12.

57. Blumberg and Gottlieb, *War on Waste*, p. 228.

58. Ecology Center, *Report*, p. 15.

59. Cited in ibid., p. 16.

60. GrassRoots Recycling Network, *Wasting and Recycling in the United States 2000*, p. 14.

61. GrassRoots Recycling Network, Taxpayers for Common Sense, Materials Efficiency Project, and Friends of the Earth, *Welfare for Waste: How Federal Taxpayer Subsidies Waste Resources and Discourage Recycling* (Athens, Ga., 1999).

62. On the procurement laws that do exist, see Seldman, "Recycling—History in the United States," pp. 2359–60.

63. On annual waste costs, see Neil Seldman, "The Fourth and Final Solid Waste Management Paradigm and the End of Integrated Solid Waste Management" (unpublished, Institute for Local Self-Reliance, 2004). On percentage of waste, see GrassRoots Recycling Network, *Wasting and Recycling in the United States 2000*, p. 5.

64. Ibid., p. 22.

65. Report cited in Peter Anderson, Brenda Platt, and Neil Seldman, "Fighting Waste Industry Consolidation with Ownership of Recycling Facilities," *Facts to Act On* #42, Institute for Local Self-Reliance, Nov. 8, 2002.

66. Quoted in Beth Baker, "Curbing Recycling Revisionists," *Environmental Action* 27, no. 2, Summer 1995, p. 31.
67. Ibid.
68. GrassRoots Recycling Network, *Wasting and Recycling in the United States 2000*, pp. 24–25.
69. Ibid., p. 13.

Chapter 8

1. Barretti Carting Corporation worker quoted in Rick Cowan and Douglas Century, *Takedown: The Fall of the Last Mafia Empire* (New York: G.P. Putnam's Sons, 2002), p. 91.
2. According to WMI's founder, "Our approach was to expand in the Sunbelt first and enter the northern part of the United States only after the southern states became saturated." See Donald L. Sexton, "Wayne Huizenga: Entrepreneur and Wealth Creator," *Academy of Management Executive* 15, no. 1 (Feb. 1, 2001).
3. "Corporate Profile: USA Waste Services Inc.," Environmental Background Information Center, see www.ebic.org/pubs/usa.html. Also see Alan A. Block, "Environmental Crime and Pollution: Wasteful Reflections," *Social Justice* 61 (March 22, 2002), electronic version.
4. According to Neil Seldman of the Institute for Local Self Reliance, the large waste corporations actually lobbied for Subtitle D and the 1993 EPA rules because they knew it would force their competition out of the market. From interview, Jan. 20, 2005. Charlie Cray, a Greenpeace Toxics Campaigner, has explained that companies like WMI support "stricter pollution control regulations because these tend to generate more business for the company without really going to the source of the problem . . . [and] stricter landfill laws which squeeze out their competitors—smaller companies, nonprofit community-based programs, and publicly-owned facilities—who cannot meet the capitalization requirements of the new laws. This process enhances monopolization of the waste collection, processing and disposal industry, and, in turn, helps . . . thwart attempts to implement long-term solutions." Quoted in "Greenwash Awards," CorpWatch, Jan. 1, 1997, see www.corpwatch.org/campaigns/PCD.jsp?articleid=4370.
5. The "hub and spoke" model, from an interview with Peter Holsclaw of the San Francisco Department of Sanitation, March 1999. WMI founder Huizenga explained that buying existing landfills was a "key criterion," since "getting a permit for a disposal site is a difficult process." Quoted in Sexton, "Wayne Huizenga."

6. "Corporate Profile: Waste Management Incorporated," Environmental Background Information Center, www.ebic.org/pubs/wmx.html. Also see Block, "Environmental Crime and Pollution."

7. "Corporate Profile: Browning-Ferris Industries," Environmental Background Information Center, see www.ebic.org/pubs/BFIAlliedprof.html. Also see Block, "Environmental Crime and Pollution."

8. "Corporate Profile: Browning-Ferris Industries."

9. Based in Scottsdale, Arizona, Allied is the country's second largest solid waste management company, with 10 million commercial and household customers in over a 100 national markets in thirty-eight states (see the Allied Waste website, investor.awin.com/phoenix.zhtml?c=74587&p=IROL-index).

10. On WMI's history, see Timothy Jacobson, *Waste Management: An American Corporate Success Story* (Washington, D.C.: Gateway Business Books, 1993), pp. 60–78, 86–94, 113. Also see Ann M. Gynn and Cheryl A. McMullen, "Waste Changes Color with Times," *Waste News* 7, no. 1 (May 14, 2001). BFI had already successfully gone public in 1968.

11. Peter Montague, "Commercial Hazardous Waste Landfills," *Rachel's Environment and Health Weekly* no. 180 (May 9, 1990), electronic version; WMI website, see www.wm.com/Templates/FAC2948/index.asp.

12. "Waste Age 100," *Waste Age* (June 1, 2004), electronic version; see www.wasteage.com. More recent estimates put garbage industry revenues at $70 billion.

13. Jeff Bailey, "Too Good to Refuse: Browning-Ferris Bucks Mob," *Wall Street Journal*, Nov. 8, 1993.

14. Bailey, "Too Good to Refuse." On New York producing 5 percent of U.S. commercial waste, see Cowan and Century, *Takedown*, p. 15.

15. Lee Linder, "Waste Industry 'Open' to Mob," Associated Press, Nov. 17, 1989; Block, "Environmental Crime and Pollution." According to Block, Mafia-style codes were in place in Chicago through the 1960s and into the 1970s, in a system called the "Chicago rules" that essentially created a cartel there as well.

16. Having already taken control of unionized waiters via corrupt channels in the Hotel and Restaurant Employees International Alliance, Jewish Mafioso Dutch Schultz started the Metropolitan Restaurant and Cafeteria Owners Association to leverage dues and fees to ensure labor tranquillity. Louis Lepke, another powerful Jewish mobster, understood that this formula could be applied to any industry. He focused on trucking, recognizing that the timely movement of commodities was key to economic health. See Cowan and Century, *Takedown*, pp. 143–46.

17. Ibid., p. 14.

18. According to the *New York Daily News*, reporting on a 1957 Senate inquiry into New York's Mobbed-up waste trade: "Under this code, if a customer moves from a 'stop,' the garbage collector who had him claims that 'stop' as his 'cousin' and no one can take it from him. If someone 'jumped' a 'stop'—took over a 'property right'—he could be fined $100 for every $10 the 'stop' brought in." Quoted in Cowan and Century, *Takedown*, pp. 33–34.
19. One extreme case of the Mafia cartel's brutality took place on the outskirts of New York City in suburban Long Island, which had its own Mafia-connected garbage cartel. After independent carter Robert Kubecka and his brother-in-law Donald Barstow agreed to help the authorities spy on Long Island's cartel, both men were lethally gunned down in their company's office. See Steve Wick, "Finally Caught," *Newsday*, Jan. 29, 2003; Steve Wick, "Used and Left Unprotected," *Newsday*, Dec. 23, 2001; Tom Renner and Michael Slackman, "They Defied the Mob," *Newsday*, Sept. 24, 1989.
20. All figures from 1993. See Cowan and Century, *Takedown*, p. 174.
21. On the $500 million overcharge, see Philip Angell, "Cleaning Up New York," *Infrastructure Finance* 6, no. 4 (May 1, 1997). It is documented that Teamsters Local 813, among other unions, worked with the cartel. See Cowan and Century, *Takedown*, pp. 70–76.
22. Cowan and Century, *Takedown*, p. 91.
23. Bailey, "Too Good to Refuse."
24. Ibid.
25. Bailey, "Too Good to Refuse"; Robin Kamen, "NY Trash Compaction: Carters Rush to Merge," *Crain's New York Business*, May 23, 1994; Angell, "Cleaning Up New York."
26. Angell, "Cleaning Up New York."
27. Bailey, "Too Good to Refuse."
28. Gary Lewi, quoted in Cowan and Century, *Takedown*, p. 174.
29. Cowan and Century, *Takedown*, pp. 175–78, 321; Angell, "Cleaning Up New York." The undercover agent who did the bulk of the work was an NYPD investigator, Rick Cowan.
30. Cowan and Century, *Takedown*, pp. 274, 314–15.
31. Jeff Bailey, "ReSource NE, Big NY Waste-Handler, in Merger Talks," *Wall Street Journal*, Nov. 27, 1995; Angell, "Cleaning Up New York"; Cowan and Century, *Takedown*, pp. 319–20, 337.
32. Jeff Bailey, "Waste Hauler in New York Looks to Deal," *Asian Wall Street Journal*, Nov. 27, 1995.

33. Steve Daniels, "Competition Comes to New York," *Waste News*, April 21, 1997.
34. Ibid.
35. Steve Daniels, "Good News, Bad News: Fresh Kills' Demise Means New Opportunities for Some, Worries for Others," *Waste News*, April 21, 1997.
36. Martin, "From the Many to the Few." A class action lawsuit was filed by customers in the late 1980s against both WMI and BFI for egregious violations of Title 1 of the Sherman Antitrust Act in eleven U.S. states and several Canadian provinces. The plaintiffs alleged that the firms routinely engaged in bid rigging, bribery of political officials, and mutual agreements to not solicit each other's customers. In 1990, WMI paid more than $30 million while BFI paid $19 million to settle the case out of court. See Block, "Environmental Crime and Pollution."
37. "Corporate Profile: Waste Management Incorporated." In the late 1990s, WMI was rocked by a series of accounting scandals due to more than a billion dollars in inflated earnings. See Mark Babineck, "Once-Tattered Trash Giant Emerges from Accounting Scandal," Associated Press, May 4, 2002.
38. Cowan and Century, *Takedown*, p. 176.
39. BFI employee quoted in ibid., p. 177.
40. Daniels, "Competition Comes to New York."
41. Kamen, "NY Trash Compaction."
42. Although the nationals' fees fell beneath inflated cartel prices, frequently they also dipped significantly lower than fair market rate. An analyst with Solomon Smith Barney noted, "Right now, probably, none of them [the nationals] are making money in any kind of material fashion. . . . At the same time, I would doubt that New York is a material drag." See Daniels, "Competition Comes to New York." The goal of predatory pricing was to weaken and drive out competition. Not long after BFI came to town, the firm started low-balling bids, and "as new bids go out, prices are expected to fall dramatically and lower the value of the hauling companies," easing acquisition. See Kamen, "NY Trash Compaction."
43. Daniels, "Competition Comes to New York."
44. Bailey, "Too Good to Refuse."
45. Philip Lentz, "Cartload of Mergers Brings New Firms to Trash Business," *Crain's New York Business*, March 15, 1999.
46. Block, "Environmental Crime and Pollution"; Lentz, "Cartload of Mergers."
47. Philip Lentz, "Carter Tightens Grip on Trash Biz," *Crain's New York Business*, Oct. 12, 1998.
48. In February 2001, there were 150 licensed haulers in New York (80 other haulers were waiting for approval from the Trade Waste Commission), as compared

with nearly 500 local carters in 1988. See "Should New York City Abolish Its Commercial Waste Hauling Rate Cap?" *Waste News* 6, no. 37 (Feb. 12, 2001). The surge in prices was largely due to the closing of Fresh Kills Landfill, discussed in the next section. See Eric Lipton, "City Seeks Ideas as Trash Costs Dwarf Estimate," *New York Times*, Dec. 2, 2003.

49. Daniels, "Good News, Bad News."

50. Cerrell Associates report for the California Waste Management Board, cited in Eddie J. Girdner and Jack Smith, *Killing Me Softly: Toxic Waste, Corporate Profit, and the Struggle for Environmental Justice* (New York: Monthly Review Press, 2002), p. 53. Also see Louis Blumberg and Robert Gottlieb, *War on Waste: Can America Win Its Battle with Garbage?* (Washington, D.C.: Island Press, 1989), p. 59.

51. Amy Waldman, "Trash Giant Skirts Conditions Set for Bronx Station, Critics Say," *New York Times Abstracts*, Aug. 24, 1999; Steve Daniels, "WMI Captures 3-Year Bronx Disposal Contract," *Waste News*, June 16, 1997.

52. Jennifer Weil, "WTC Air Grant Deadline Near," *New York Daily News*, Nov. 27, 2002.

53. Quoted in Amy Waldman, "In South Bronx, a Bitter Split," *New York Times Abstracts*, Sept. 29, 1999.

54. Quoted in John McQuaid, "Landfill Is Fertile Ground for Political Fight," *New Orleans Times-Picayune*, May 21, 2000.

55. Waldman, "Trash Giant Skirts Conditions."

56. On daily tonnage, see Waldman, "Trash Giant Skirts Conditions" and "In South Bronx, a Bitter Split." The 40 percent figure relates specifically to the Hunt's Point section of the South Bronx; see McQuaid, "Landfill Is Fertile Ground."

57. Lipton, "City Seeks Ideas"; Michael Clancy, "Buried in Trash," *AM New York* 1, no. 38 (Dec. 3, 2003).

58. Valerie Burgher, "Breathing Lessons," *Village Voice*, April 29, 1997. Also see Weil, "WTC Air Grant."

59. "Soot Is Cited as Big Factor in Global Warming," *New York Times*, Dec. 25, 2003. Also see David Adam, "There Goes the Sun," *Manchester Guardian Weekly*, Dec. 25–31, 2003.

60. In 1998 Pennsylvania was the country's biggest importer of trash at more than 6 million tons annually. At the time, Virginia, the country's number two importer, took in almost 3 million tons, most of which got sent to one of seven new megafills. See Eric Lipton, "As Imported Garbage Piles Up, So Do Worries," *Washington* Post, Nov. 12, 1998.

61. According to the U.S. Census. See Lipton, "Imported Garbage."

62. After testing, it was revealed that the ash contained high levels of lead, cadmium and dioxins with lower but still alarming levels of chromium and arsenic. See Center for Investigative Reporting and Moyers, *Global Dumping Ground*, pp. 19–30.
63. The full memo is reproduced on the *Counterpunch* website, www.counterpunch.org/summers.html. Summers later recanted this statement, explaining that it was meant to be ironic.
64. From an interview with Dave Williamson, Berkeley Ecology Center, May 2000.
65. Karl Schoenberger, "E-Waste Ignored in India," *San Jose Mercury News*, Dec. 28, 2003.
66. Beverly Burmeier, "Happy Endings for Cast-Off PC's," *Christian Science Monitor*, April 21, 2003; "Americans as Consumers of and Contributors to World Resources," *Research Alert*20, no. 13 (July 5, 2002).
67. "Exporting Harm: The High-Tech Trashing of Asia," Basel Action Network and Silicon Valley Toxics Coalition (Seattle, Feb. 25, 2002), p. 7.
68. Burmeier, "Happy Endings."
69. Henry Norr, "Drowning in E-Waste," *San Francisco Chronicle*, May 27, 2001.
70. Basel Action Network and Silicon Valley Toxics Coalition, "Exporting Harm," p. 7.
71. Kyung M. Song, "Toxic High-Tech Waste Flows to Asia," *Seattle Times*, Feb. 25, 2002.
72. Ibid.
73. Katherine Stapp, "Cheap Cell Phones Increase Piles of E-Waste in South," Inter Press Service, May 17, 2002.
74. Schoenberger, "E-Waste Ignored in India."
75. Jon E. Hilsenrath, "Beijing Strikes Gold with U.S. Recycling," *Asian Wall Street Journal*, April 9, 2003.
76. Center for Investigative Reporting and Bill Moyers, *Global Dumping Ground*, pp. 7, 9.
77. Quoted in Neil Smith, *The New Urban Frontier: Gentrification and the Revanchist City* (London: Routledge, 1996), p. 28 (italics in original).

Chapter 9

1. Based on visits to the Batcave and multiple interviews with Tim Krupnick, 2001–2002. Also see Tim Krupnick, "The Urban Wilds Project," *Permaculture Activist* 6, no. 45 (March 2001), pp. 63–65.
2. Daniel Imhoff, "Thinking Outside of the Box," *Whole Earth* (Winter 2002), p. 9.

3. Caycee Cullen, "The Jar," instruction 'zine, 2003.

4. Tina Kelley, "One Sock, with Hole? I'll Take It," *New York Times*, March 16, 2004. Also see Freecycle website, www.freecycle.org.

5. From the documentary film *Gone Tomorrow*.

6. George Monbiot, "Fuel for Nought," *Guardian Weekly*, Dec. 3, 2004. It would be naïve to think the petroleum sector would give up the booming plastics trade just because new synthetics weren't made with petrochemicals.

7. Imhoff, "Thinking Outside of the Box."

8. McDonough and Braungart, *Cradle to Cradle*, p. 151.

9. According to urban geographer Richard Walker, "Capital's only relation to nature is ultimately its relation to accumulation. But the technological power is so great now that the law of accumulation can mean a law of destruction of the earth and of life on earth at a scale never before seen. . . Social regulation is an inherent part, a necessary part of social life under modern production." From the documentary film *Gone Tomorrow*.

10. "Setting the Record Straight," available from the Container Recycling Institute website, www.container-recycling.org/plasfact/PETstraight.htm.

11. Des King, "Calling the Shots," *Packaging Today International* 25, no. 9 (Sept. 2003).

12. Sara Bloom, "How Is Germany Dealing with Its Packaging Waste?" *Whole Earth* (Winter 2002), pp. 23–25.

13. From an interview with Frank Ackerman, April 22, 2004.

14. Imhoff, "Thinking Outside of the Box," p. 14; Jim Motavalli, "Zero Waste," *E the Environmental Magazine* 12, no. 2 (March/April 2001), electronic version.

15. "Germany's Green Dot System Challenged," *UK: Environment News*, March 30, 2004; "Packaging Recycling for a Better Climate," *The OECD Observer*, Dec. 2002.

16. Thanks to this guideline, the government was able to rein in producers who in recent years marketed almost twice as many single-use containers as was allowed. Otto Pohl, "Sales Slow as Germans Pile Up Empties," *New York Times*, March 5, 2003.

17. Motavalli, "Zero Waste."

18. Bloom, "How Is Germany Dealing?" p. 24.

19. As of 2003, the annual per capita consumption for packaging in Germany was €319, in the United Kingdom it was €234, and in Russia (whose population exceeds Germany's by 77 percent) the figure was a miniscule €19. See King, "Calling the Shots."

20. All green dot packaging that does not go to recycling plants is burned or buried

outright. See Rob Edwards, "Waste Not, Want Not: Outside Every German Home Stands a Multicolored Set of Dustbins," *New Scientist* (Dec. 23, 1995), electronic version.

21. Ibid.

22. San Francisco's program is dubbed "Fantastic Three" and is operated by the refuse company Norcal. See Kim A. O'Connell, "San Francisco Giant," *Waste Age* 34, no. 12 (Dec. 1, 2003).

23. This point came out of a correspondence with my editor, Liza Featherstone.

24. Brenda Platt and Doug Rowe, *Reduce, Reuse, Refill!* (Washington, D.C.: Institute for Local Self-Reliance, April 2002), produced under a joint project with the GrassRoots Recycling Network, p. 27.

25. Ibid., pp. 1–2.

26. Ibid., p. 1.

27. Ibid., p. 33.

28. Ibid., pp. 5–9.

29. Ibid., pp. 14–15.

30. Ibid., p. 33.

31. If companies are not trying to centralize production and distribution, as the giant firms like Coca-Cola have done, then refillables can actually lower costs. See ibid., p. 25.

32. Paul Millbank, "Aluminum Recycling Vital to Global Supply Chain," *Aluminum International Today* 16, no. 5 (Sept. 1, 2004); "Bottle Battle," *Economist Intelligence Unit—Country Monitor* (Nov. 3, 2003).

33. Platt and Rowe, *Reduce, Reuse, Refill!* pp. 27–29, 40–41.

34. Eric Lombardi, "Beyond Recycling! Zero Waste . . . Or Darn Near"; see GrassRoots Recycling Network website, www.grrn.org/zerowaste/articles/biocycle_zw_commentary.html.

35. GrassRoots Recycling Network website, www.grrn.org/zerowaste/zerowaste_faq.html.

36. Ibid.

37. Bill Sheehan and Daniel Knapp, "Zeroing In on Zero Waste," see GrassRoots Recycling Network website, www.grrn.org/zerowaste/articles/zeroing_in.html.

38. Kate Soper, "Waste Matters," *Capitalism Nature Socialism* 14 (2), no. 54 (June 2003), pp. 131–32. As noted in previous chapters, for every ton of household waste, more than 70 tons of manufacturing wastes are produced. See GrassRoots Recycling Network, *Wasting and Recycling in the United States 2000*, p. 13.

39. According to the GrassRoots Recycling Network, *Wasting and Recycling in the*

United States 2000: "On a per-ton basis, sorting and processing recyclables alone sustains ten times more jobs than landfilling or incineration." Cited in Eric Lombardi, "Beyond Recycling! Zero Waste . . . Or Darn Near."

40. GrassRoots Recycling Network, *Wasting and Recycling in the United States 2000*, p. 27.

41. Ibid., p. 28.

42. Gary Liss, "Zero Waste?" GrassRoots Recycling Network website, www.grrn.org/zerowaste/articles/whatiszw.html. For more on jobs, see Paula DiPerna, "Mean Green Job Machine," *Nation* 278, no. 20 (May 24, 2004), p. 7.

43. One example: Gabriela Zamorano explained in a November 2003 conversation that an indigenous group that she worked with in Oaxaca, Mexico, has no word for garbage in their native language.

44. "Drowning in a Tide of Discarded Packaging," *Guardian* (London), March 9, 2002.

45. Marx, *Capital*, p. 1:279.

46. Joel Kovel, *The Enemy of Nature: The End of Capitalism or The End of The World?* (London: Zed Press, 2002), p. 155.

47. On the undemocratic nature of the use of natural resources, see Soper, "Waste Matters"; Commoner, *Making Peace with the Planet*.

48. "The U.S. Labor Force in the New Economy," *Social Education* 68, no. 2 (March 1, 2004), electronic version.

49. "USA Risk: Macroeconomic Risk," *Economist Intelligence Unit—Riskwire*, no. 101 (July 7, 2003), electronic version.

50. Statistical evidence shows that layoffs are more frequent now, whether the economy is on the upswing or in the trough of a downturn. See Louis Uchitelle, "Layoff Rate at 8.7%, Highest Since 80's," *New York Times*, Aug. 2, 2004.

51. Leslie Sklair, *The Transnational Capitalist Class* (Oxford: Blackwell, 2001), p. 207.

Index